Biosensor and Chemical Sensor Technology

ACS SYMPOSIUM SERIES 613

Biosensor and Chemical Sensor Technology

Process Monitoring and Control

Kim R. Rogers, EDITOR
U.S. Environmental Protection Agency

Ashok Mulchandani, EDITOR
University of California—Riverside

Weichang Zhou, EDITOR
Merck & Co., Inc.

Developed from two symposia sponsored
by the Division of Biochemical Technology
and the Biochemical Secretariat
at the 209th National Meeting
of the American Chemical Society,
Anaheim, California,
April 2–6, 1995

American Chemical Society, Washington, DC 1995

Library of Congress Cataloging-in-Publication Data

Biosensor and chemical sensor technology: process monitoring and control / Kim R. Rogers, editor; Ashok Mulchandani, editor; Weichang Zhou, editor.

p. cm.—(ACS symposium series, ISSN 0097–6156; 613)

"Developed from a symposium sponsored by the Division of Biochemical Technology and the Biochemical Secretariat at the 209th National Meeting of the American Chemical Society, Anaheim, California, April 2–6, 1995."

Includes bibliographical references and indexes.

ISBN 0–8412–3330–6

1. Biosensors—Congresses. 2. Chemical detectors—Congresses. 3. Biotechnological process monitoring—Congresses. 4. Biotechnological process control—Congresses.

I. Rogers, Kim R., 1956– . II. Mulchandani, Ashok, 1956– III. Zhou, Weichang, 1963– . IV. American Chemical Society. Division of Biochemical Technology. V. Series.

R857.B54B534 1995
660′.6′0287—dc20 95–39897
CIP

This book is printed on acid-free, recycled paper.

PRINTED IN THE UNITED STATES OF AMERICA

1995 Advisory Board

ACS Symposium Series

Foreword

THE ACS SYMPOSIUM SERIES was first published in 1974 to provide a mechanism for publishing symposia quickly in book form. The purpose of this series is to publish comprehensive books developed from symposia, which are usually "snapshots in time" of the current research being done on a topic, plus some review material on the topic. For this reason, it is necessary that the papers be published as quickly as possible.

Before a symposium-based book is put under contract, the proposed table of contents is reviewed for appropriateness to the topic and for comprehensiveness of the collection. Some papers are excluded at this point, and others are added to round out the scope of the volume. In addition, a draft of each paper is peer-reviewed prior to final acceptance or rejection. This anonymous review process is supervised by the organizer(s) of the symposium, who become the editor(s) of the book. The authors then revise their papers according to the recommendations of both the reviewers and the editors, prepare camera-ready copy, and submit the final papers to the editors, who check that all necessary revisions have been made.

As a rule, only original research papers and original review papers are included in the volumes. Verbatim reproductions of previously published papers are not accepted.

Contents

Preface xi

BIOSENSOR METHODS

1. **Biomolecular Sensing for Bioprocess and Environmental Monitoring Applications** 2
   Kim R. Rogers and Ashok Mulchandani

2. **Use of an Acoustic Wave Device as a Liquid Chromatography Detector** 9
   May Tom-Moy, Thomas P. Doherty, Richard L. Baer, and Darlene Spira-Solomon

3. **Immunosensors for Detection of Chemical Mixtures: Antibody Affinities, Selectivities, and Cloning** 19
   Mohyee Eldefrawi, Amira Eldefrawi, Jeremy Wright, Peter Emanuel, James Valdes, and Kim R. Rogers

4. **Adaptation of a Fiber-Optic Biosensor for Use in Environmental Monitoring** 33
   Mark D. Pease, Lisa Shriver-Lake, and Frances S. Ligler

5. **A New Method for the Detection and Measurement of Polyaromatic Carcinogens and Related Compounds by DNA Intercalation** 44
   John J. Horvath, Manana Gueguetchkeri, Adarsh Gupta, Devi Penumatchu, and Howard H. Weetall

6. **Chemically Modified Electrode for Hydrogen Peroxide Measurement by Reduction at Low Potential** 61
   Ashok Mulchandani and Lisa C. Barrows

7. **Enzyme Sensors for Subnanomolar Concentrations** 70
   Frieder W. Scheller, Alexander Makower, Andrey L. Ghindilis, Frank F. Bier, Eva Förster, Ulla Wollenberger, Christian Bauer, Burkhard Micheel, Dorothea Pfeiffer, Jan Szeponik, Norbert Michael, and H. Kaden

8. **Electroenzymatic Sensing of Fructose Using Fructose Dehydrogenase Immobilized in a Self-Assembled Monolayer on Gold** **82**
K. T. Kinnear and H. G. Monbouquette

BIOPROCESS MONITORING AND CONTROL

9. **Recent Advances in Bioprocess Monitoring and Control** **88**
Weichang Zhou and Ashok Mulchandani

10. **Optical Measurement of Bioprocess and Clinical Analytes Using Lifetime-Based Phase Fluorimetry** **99**
Shabbir B. Bambot, Joseph R. Lakowicz, Jeffrey Sipior, Gary Carter, and Govind Rao

11. **Biosensor for On-Line Monitoring of Penicillin During Its Production by Fermentation** **110**
Canh Tran-Minh and Helmut Meier

12. **Selective Measurement of Glutamine and Asparagine in Aqueous Media by Near-Infrared Spectroscopy** **116**
Xiangji Zhou, Hoeil Chung, Mark A. Arnold, Martin Rhiel, and David W. Murhammer

13. **An Expert System for the Supervision of a Multichannel Flow Injection Analysis System** **133**
B. Hitzmann, R. Gomersall, J. Brandt, and A. van Putten

14. **Hybrid Process Modeling for Advanced Process State Estimation, Prediction, and Control Exemplified in a Production-Scale Mammalian Cell Culture** **144**
M. Dors, R. Simutis, and A. Lübbert

15. **Optimization of an *Escherichia coli* Fed-Batch Fermentation Using a Turbidity Measurement System** **155**
Yinliang Chen, Alahari Arunakumari, Glen Hart, and Julia Cino

16. **The Automation of Two Flow-Injection Immunoassays Using a Flexible Software System** **165**
B. Hitzmann, B. Schulze, M. Reinecke, and T. Scheper

INDEXES

Author Index 178

Affiliation Index 178

Subject Index 179

Preface

ANALYTICAL TECHNOLOGIES ARE AT THE HEART of any process control application. Sensitive, selective, reliable, and robust analytical tools are necessary for on-line monitoring and control applications in the environmental, clinical, and bioprocess areas. In response to this need, a variety of analytical methods have been demonstrated for these complex monitoring problems. Although many of these methods are based on diverse transducers (e.g., optical, electrochemical, and acoustic), each has been applied to the identification and measurement of specific compounds either through direct interrogation of the sample or through the interface of a biological recognition element (biosensors).

This volume presents a cross section of recent advances in the development of novel chemical and biochemical sensors for on-line monitoring and control applications in the environmental, clinical, and bioprocess areas. These chapters illustrate how many of the key challenges for continuous monitoring are being addressed. The methods discussed include optical techniques ranging from near-infrared spectroscopy to lifetime-based phase fluorometry; biosensors ranging from optical immunosensors to enzyme-electrodes; as well as electrochemical, acoustic, and plasmon resonance techniques.

The chapters in this book have been divided into two sections, "Biosensor Methods" (Chapter 1–8) and "Bioprocess Monitoring and Control" (Chapters 9–16). Chapters 1–8 discuss the development of methods that rely on biomolecular recognition for detection of target analytes for potential environmental and bioprocess monitoring applications. Recognition elements primarily include enzymes, antibodies, microorganisms, and DNA. These elements have been interfaced to electrochemical, optical, and acoustic sensors. The range of potential analytes is broad, and many of these methods can be modified to accommodate the measurement of additional analytes through the introduction of new enzymes or antibodies.

Chapters 9–16 cover the application of optical and biosensor methods for on-line monitoring and control of bioprocesses. Topics include the use of expert systems for controlling flow injection analysis systems and hybrid process modeling for bioprocess control.

Each of the chapters, except section introduction Chapters 1 and 10, were presented in the two symposia, Biomolecular Sensing for Process Control Applications and Advances in Monitoring and Control of

Bioprocesses, organized by the Biotechnology Secretariat in co-sponsorship with the Division of Biochemical Technology.

Acknowledgments

We are fortunate to have assembled contributions from world-class authorities in this field. We sincerely thank all the participants, who not only made these symposia successful but also have made an important contribution to the literature through this book. We thank the Biotechnology Secretariat and the Division of Biochemical Technology for coordination of the two symposia, the reviewers who provided excellent comments to the editors and contributors in this book, and the supportive staff of ACS Books. The gracious support of our families is most warmly acknowledged.

KIM R. ROGERS
Characterization Research Division
National Exposure Research Laboratory
U.S. Environmental Protection Agency
P.O. Box 93478
Las Vegas, NV 89193–3478

ASHOK MULCHANDANI
Chemical Engineering Department
College of Engineering
University of California
Riverside, CA 92521

WEICHANG ZHOU
Bioprocess Research & Development Department
Merck Research Laboratories
Merck & Co., Inc.
P.O. Box 2000
Rahway, NJ 07065

July 24, 1995

Biosensor Methods

Chapter 1

Biomolecular Sensing for Bioprocess and Environmental Monitoring Applications

Kim R. Rogers[1] and Ashok Mulchandani[2]

[1]**National Exposure Research Laboratory, Characterization Research Division, U.S. Environmental Protection Agency, P.O. Box 93478, Las Vegas, NV 89193–3478**
[2]**Chemical Engineering Department, College of Engineering, University of California, Riverside, CA 92521**

Biomolecular recognition is being increasingly employed as the basis for a variety of analytical methods such as biosensors. The sensitivity, selectivity, and format versatility inherent in these methods may allow them to be adapted to solving a number of analytical problems. Although these methods are highly diverse with respect to their mechanisms of biomolecular recognition and mechanisms of signal transduction, they can be viewed together as providing analytical techniques for the identification and measurement of compounds critical to potential monitoring applications. This chapter will provide a brief introduction to recent advances in biomolecular sensing as applied to monitoring of bioprocesses and the environment.

Biomolecular recognition provides the basis for chemical process monitoring and control for the most complex and sophisticated systems known--living organisms. Nevertheless, although analytical methods which rely on biological recognition show considerable promise, there are a great number of challenges and limitations in exploiting these systems for potential process monitoring applications. This is particularly true in the bioprocess and environmental monitoring areas, each having unique and specific requirements.

Although potential bioprocess monitoring applications (such as fermentations) and environmental applications (such as remediation of hazardous waste sites or waste water monitoring at water treatment facilities) are substantially different, certain requirements for monitoring these processes are similar. For example, in each case, analytical methods using biomolecular recognition may allow the identification and quantitation of specific compounds; classes of compounds; macromolecules (e.g., antibodies or other proteins); or microorganisms which are critical to the process which one wishes to control (1, 2). In the application of these devices and techniques to environmental monitoring or on-line monitoring for control of a bioprocess, three essential components must be present, i.e. (i) an analytical technique or device (e.g. biosensors); (ii) a suitable configuration employing

0097–6156/95/0613–0002$12.00/0

an *in situ* or *ex situ* arrangement for contacting the fermentation medium or environmental matrix with the device or technique; and (iii) a reporting or control system in order to employ a control strategy. Methods which will allow real-time and *in situ* operation could, in certain circumstances, provide cost-effective solutions to difficult and persistent problems in both bioprocess and environmental areas.

Potential application areas for the use of biomolecular recognition-based methods (such as biosensors) in process monitoring and control are expansive. For the purpose of this chapter, we will focus on biosensing concepts and applications for the detection of substrates and products of fermentation processes and the monitoring of environmental pollutants.

Biomolecular Recognition

Biomolecular recognition depends on the binding of ions, small molecular weight organics, and biological macromolecules to biological receptors. Mechanisms for detecting and measuring the recognition event depend primarily on the action of the biological receptor. These receptors can be grouped functionally as biocatalytic (e.g., enzymes), bioaffinity (e.g., receptors, antibodies, and DNA), or microbial (e.g., bacteria) in nature. When these biomolecular recognition elements are directly coupled to signal transducers (e.g., electrochemical, optical, or acoustic), they can generate an electrical signal proportional to the target analyte concentration and are typically referred to as biosensors.

Theoretically, any biological recognition element can be interfaced to any of these types of transducers, provided an appropriate reaction product or analyte probe can be devised and measured. There are, however, a number of technical and practical issues which must be considered in the development of biosensors for environmental and bioprocess monitoring applications. In addition to a sensitive and specific response to the target analyte, biosensors which are most likely to find commercial success for these application areas, will require simple and inexpensive configurations which lend themselves easily to miniaturization and manufacturing techniques.

One of the greatest potential advantages of biosensors for these applications is the versatility, not only in the biorecognition elements, but also in the operational formats. For example, biosensor methods for the immunochemical detection of herbicides have been reported using reversible (3), regenerable (4), and disposable (5) biosensor formats. An important feature which must be considered is whether the mechanism of signal transduction allows the *in situ* or *ex situ* contact of the medium. This is particularly important for use with bioreactors which require strict adherence to sterilization protocols. Another consideration is the requirement for additional substrates, cofactors, or labeled analyte probes. This can present technical problems for potential *in situ* applications, especially where continuous operation is required or a sterile environment must be maintained.

Monitoring for process control applications

Recent advances in the field of biotechnology have produced a wealth of novel products. However, due to high production costs a number of these products are not commercially viable. The production costs can be lowered by improving the overall bioprocess efficiency. This will require development of control strategies to maintain concentrations of key metabolites that support the biological growth process at its optimal level. To date, control strategies for bioprocesses relied have on predictive models. However, the generic application of predictive models faces severe limitations; particularly for recombinant organisms and tissue cell cultures, where a detailed understanding of their metabolic pathways is presently lacking. Preferable to the model based control strategy is a feedback control system based on direct measurement, by sensors, of key control parameters.

The current state of sensor technology for bioprocess applications permits the measurement and control of dissolved oxygen, pH, temperature, agitation and foam in a bioreactor. Recent advances in sensors technology have lead to on-line determination of biomass through *in situ* optical density and fluorometric probes (6). Oxygen uptake or carbon dioxide evolution rates during cell cultivation have been followed using mass balances and off-gas analysis (7). Several novel approaches to monitor volatile organic compounds have also been considered (8, 9) and one such sensor for methanol was commercialized. However, no reliable technique exists to carry out real time analysis of non-volatile compounds in bioprocesses. A number of non-volatile products continue to be measured off-line using HPLC and other wet chemistry techniques. HPLC and other wet chemistry techniques integrated to flow injection analysis (FIA) have also been demonstrated for on-line monitoring of bioprocess variables. These methods, however, are time consuming, labor intensive and expensive; wet chemistry techniques may require sample preparation and in the case of HPLC requires a long time to achieve separation. This decreases the usefulness of the measurement.

Because of the advantages of selectivity, sensitivity, and versatility offered by the biological recognition element, analytical techniques based on biosensors are ideal for the direct (without sample preparation) measurement of non-volatile compounds in a complex bioprocess milieu. A number of biosensors have been reported for detection of compounds (such as ethanol, lactate, urea, amino acids, and sugars) (10) associated with bioprocesses. Nutrients, intermediates, and products in bioprocesses have been monitored selectively using enzyme- and affinity-based biosensors. However, there are limitations when it comes to the application of these devices for *in situ* monitoring. Because biosensors, in general, (i) cannot be sterilized, (ii) function within a limited range of analyte concentrations, and (iii) show optimum operating conditions which often differ from that required for bioprocesses, they are not suitable for *in situ* monitoring.

Biosensors, however, are well suited for *ex situ* monitoring using FIA. The incorporation of biosensors in FIA offers many advantages such as: (i) preconditioning of analyte samples for optimal analytical conditions (for example, pH, buffer capacity) before analysis, (ii) recalibration of sensor to counter the drift, (iii) replacement of non-functional or poorly functioning parts of the analytical system such as an inactivated biological element, (iv) multi-component monitoring with one sample, and (v) short response time (typically, in the range of a few seconds to a few minutes), that are particularly beneficial for bioprocess monitoring. Because of relatively short response times, biosensor incorporated FIA methods can be considered to be quasi on-line. However, for a meaningful feedback control, it is extremely important to keep the delay from sampling to analysis as small as possible. In bioporcesses where the growth rates are extremely high, or during a critical phase of a bioprocess, even shorter intervals are necessary to enable accurate monitoring of the dynamics (11).

The aseptic sampling of analyte samples is a critical process in on-line monitoring using biosensors incorporated in FIA. *In situ* filtration probes or *ex situ* cross-flow filtration devices have been used for aseptic sampling. The advantage of these sampling systems is that they form a sterile barrier between the bioreactor and the analytical system. However, serious errors in analysis results can occur due to change in the transmembrane permeation rate and cut-off of the membrane during cultivation over a prolonged period. This is especially true for cultivation media with high protein content and reactors with low turbulence, such as mammalian cell cultures (12). It is therefore important that before any FIA incorporated biosensor is used for a process control application that a systematic investigation be carried out for several months in order to validate the analytical method.

Environmental Monitoring Applications

Analytical methods which depend on biomolecular recognition in general and more specifically biosensor methods, show the potential to complement a number of emerging screening and monitoring methods for environmental applications. Advantages shown by analytical methods which use biomolecular recognition arise primarily from the sensitivity, specificity, and versatility of the biological receptors. For example, antibody-based biosensors have been shown to selectively bind compounds ranging from small molecular weight organics (3) to genetically engineered microorganisms (13); enzyme-biosensors measure compounds of environmental concern through substrate catalysis (14) or through specific inhibition of catalytic activity by certain of these contaminants (15); bacterial-based biosensors have been reported to detect environmental toxicants (16); and DNA-based biosensors have been shown to detect compounds on the basis of their potential activity as carcinogens (17). Further, biosensors have been reported to measure a fairly broad spectrum of environmental processes and pollutants including: ammonia, non-volatile organics, heavy metals, pesticides, bacteria, and biological oxygen demand (BOD) (2). With the exception of BOD biosensors, however, none of these biosensors are commercially available for environmental applications.

Potential application areas include: laboratory screening, field screening, or continuous and *in situ* field monitoring. The cleanup of a hazardous waste site may provide some examples of the scope and kinds of analytical tasks required for environmental applications. Analytical tasks associated with site characterization primarily involve the identification of listed contaminants and mapping of the spatial distribution of the compounds of concern. The diagnostic analytical tasks are best suited to classical laboratory-based methods such as GC, GC-MS, and LC. Screening tasks to determine the spatial distribution of target analytes, however, might be accomplished most cost-effectively using laboratory or field screening methods which rely on biomolecular recognition. Although a variety of field screening methods for environmental contaminants have been developed ranging from chemical and immunochemical test kits to portable GC, biosensors using flow injection analysis (FIA) or as detectors for portable LC systems, may prove competitive for certain applications.

Once the site has been described in terms of the spatial distribution of specific pollutants, the analytical tasks associated with remediation and post-closure typically require frequent and repetitive analysis at specific locations for particular compounds of interest. Because biosensors show the potential to operate continuously at remote or *in situ* locations, these devices could be particularly well suited for this task. For example, in many cases during remediation procedures on-site, real-time monitoring may be required to prevent off-site contamination of groundwater, especially where flow patterns are quickly and dramatically altered as a result of remediation procedures such as soil excavation, treatment, and backfilling.

Another application for which biosensors may prove useful is in the post-closure groundwater monitoring. After wells have been established, monitoring must continue (in some cases mandated by law) even though the contamination has been contained, and samples are consistently returned from laboratory analysis as non-detects. In these cases, a sentinel capability which could continuously monitor for non-compliance analyte concentrations would be highly cost-effective.

Future Directions

Recent advances in biomolecular recognition and sensing techniques have been focused on a variety of fields (e.g., food analysis, agricultural chemicals, clinical diagnostics, bioprocess, and environmental monitoring). Areas for which these methods show particular promise is in solving difficult and persistent problems related to bioprocess and environmental monitoring. This is primarily because solutions to these analytical problems demand fast, reliable, cost-effective, sensitive, and specific methods which, in many cases, can be adapted to continuous and *in situ* formats. Nevertheless, before biosensors make a significant contribution to solving these problems, there exist a number of challenges which remain to be solved. These challenges fall primarily in two main areas; technical challenges associated with the operational characteristics and practical considerations related to manufacturing, marketing, and (for the environmental user) regulatory issues.

Technical challenges primarily involve, the diversity of potential target compounds (both in the bioprocess and environmental areas), the interaction of matrix compounds and conditions (i.e., pH, ionic strength, or potentially toxic compounds for environmental applications), method calibration (particularly for continuous or *in situ* applications), the requirement for reliable and low maintenance functioning over extended periods of time, and interfacing with required media, in terms of sterilization for bioprocess and multi-media (i.e., water, soil, and sludge) for environmental applications.

Practical considerations primarily involve meeting certain prerequisite requirements for commercialization such as: a viable and working technology, a substantial cost-per-assay benefit over other methods, the ability to manufacture the biosensor at a competitive cost, regulatory acceptance (EPA or FDA), a sufficient and clearly identified market, and a means to provide sales and service. Given these practical and technical obstacles, the relative paucity of commercially developed biosensors is not surprising. There is, however, sufficient evidence to suggest that with continued research, development, and validation, biosensors will provide solutions in a variety of niche applications in the bioprocess and environmental areas.

Notice

The U.S. Environmental Protection Agency (EPA), through its Office of Research and Development (ORD), funded (in part) the work involved in preparing this article. It has been subject to the Agency's peer review and has been approved for publication. The U.S. Government has the right to retain a non-exclusive, royalty-free license in and to any copyright covering this article.

Literature Cited

1. Mulchandani, A.S.;Bassi, S. *Crit. Rev. Biotech.* 1990, (in press).
2. Rogers, K.R. *Trends Anal Chem.* 1995 (in press).
3. Wong, R.B.; Anis, N.; Eldefrawi, M.E. *Anal. Chim Acta* 1993, 279, 141-147.
4. Bier, F.F.; Jockers, R.; Schmidt, R.D. Analyst 1994, 119, 437-441.
5. Oroszlan, P.; Duveneck, G.L.; Ehrat, M.; Widmer, H.M. *Sens. Actuat.* 1993, B, 11, 301-305.
6. Luong, J.H.T.; Mulchandani, A. In *Sensors for Bioprocess Control*; J.V., Twork, A.M., Yacynych, Eds.; Marcel Dekker, NY, 1990, pp 75-94.
7. Suzuki, T.; Yamane, T.; Shimizu, S. J. *Fermen. Technol.* 1986, 64, 317-326.
8. Arminger, W.B. In *Comprehensive Biotechnology: The Principles, Applications, and Regulations of Biotechnology in Industry, Agriculture, and Medicine*; M. Moo-Young, Ed.; Pergamon Press, NY, 1985, pp133-148.
9. Austin, G.D.; Sankhe, S.K.; Tsao, G.T. *Bioprocess Eng.* 1992, 7, 241-247.
10. Bilitewski, U. In *Food Biosensor Analysis*; G. Wagner, G.G. Guilbault, Eds.; Marcel Dekker, Inc., NY, 1994, pp 31-62.
11. Scheper, T.; Plotz, F.; Muller, C.; Hitzmann, B. *Trends Biotechnol.* 1994, 12, 42-46.
12. Schugerl, K. J. Biotechnol. 1993, 31, 241-256.
13. Pease, M.D.; Shriver-Lake, L.C.; Ligler, F.S. In *Recent Advances in Biosensors, Bioprocess Monitoring, and Bioprocess Control*; K.R. Rogers,; A. Mulchandani; W. Zhou, Eds.; ACS Symposium Series; American Chemical Society, Washington, D.C., 1995, (this vol).
14. Wang, J.; Yuehe, L. *Anal. Chim. Acta* 1993, 271, 53-58.
15. Rogers, K.R.; Cao, C.J.; Valdes, J.J.; Eldefrawi, A.T.; Eldefrawi, M.E. *Fundam. Appl.Toxicol.* 1991, 16, 810-820.
16. De Zwart, D.; Sloof, W. *Aquat. Toxicol.* 1983, 4, 129-138.

17. Horvath, J.J.; Gueguetchkeri, M.; Penumatchu, D.; Weetal, H.H. In *Recent Advances in Biosensors, Bioprocess Monitoing, and Bioprocess Control*; K.R. Rogers,; A. Mulchandani; W. Zhou, Eds.; ACS Symposium Series; American Chemical Society, Washington, D.C., 1995, (this vol).

RECEIVED June 30, 1995

Chapter 2

Use of an Acoustic Wave Device as a Liquid Chromatography Detector

May Tom-Moy, Thomas P. Doherty, Richard L. Baer, and Darlene Spira-Solomon

Chemical Systems Department, Hewlett-Packard Laboratories, Hewlett-Packard Company, 3500 Deer Creek Road, Palo Alto, CA 94303

We have developed a proprietary acoustic wave device which permits the detection of a specific analyte in a flowing system. By coupling specific chemistry (Protein A) to the surface of the device, the mass loading of the device by the target analyte (Human IgG) was detected as a shift in phase which was measured in real time. Using conditions which mimic a bioprocess separation for IgG, we were able to separate and detect Human IgG at 1 mg/ml and 100 ug/ml in the absence and presence of 10% Fetal Bovine Serum. Such a detector has the potential to increase productivity in process chromatography in biopharmaceutical applications.

Detectors for liquid chromatography can be classified as either universal or selective. The most common example of the former is the ultraviolet-visible (UV-Vis) detector. There are many examples of selective detectors, including electrochemical and fluorescence detectors (1-5). Their selectivity is often based on the response of a functional group that is present in some, but not all, of the components of a sample. Few, if any, of the selective detectors developed to date can be made selective enough to respond only to a single compound. There are many cases where selective detection is desirable. In bioprocess chromatography, for example, the objective is to separate the cellular product of interest from cell media in a fermentation vessel. Methods must be developed that result in the desired product eluting as a pure peak. Ideally, one would want two detectors on a system used for method development: a detector that responds to every component in the sample, and one that responds only to the desired product. With such a system, the user could determine if a given separation had achieved the goal of isolating the product.

The way users currently gather this information is by monitoring the chromatographic peaks with a UV-VIS spectrophotometer, and collecting the fractions of potential interest. These fractions are then further analyzed using any number of biochemical analyses, such as Western blotting, SDS-PAGE, ELISA's or

0097–6156/95/0613–0009$12.00/0

protein sequencing in order to quantitate the peaks and determine the analyte's identity. Such a procedure is time consuming and costly in terms of labor and production. When a new procedure is required for separating new compounds, methods development is hampered by the slow feedback of identifying the specific peaks. What is needed in this situation is a detector that produces the same information as an ELISA but in a continuous manner without the need to collect fractions and perform other manual steps. The need for fast reaction times and continuous output rules out multi-step, sandwich-type assays and most multiplication strategies.

Previous attempts to develop post column on-line detectors have been in the form of two cell reactors (1,6). The first cell is an enzyme reactor in which a specific enzyme is immobilized on a solid phase media, i.e., beads. Within this enzyme reactor cell, a reaction takes place between the separated product and the immobilized enzyme to generate another detectable product such hydrogen peroxide. The hydrogen peroxide is then detected by a second reactor cell, an electrochemical detector, which makes an amperometric determination of hydrogen peroxide. Different oxidases can be used in the enzyme reactor cell which react with sugars, acids or alcohols to generate hydrogen peroxide. Such a detection system takes advantage of the specificity of immobilized enzymes and the high sensitivity of electrochemical detection. There are however drawbacks to this method. This particular configuration requires that the eluant flow through two cells rather than one cell. The detection of the target analyte is indirect since it is based on the generation of hydrogen peroxide which is the byproduct of the reaction between the target analyte and the oxidase. Furthermore, the analyte to be measured must react with an enzyme in order to mediate a response. In other systems where enzymes are not utilized, the detection cell often incorporates an immobilized protein which captures the target compound (5). However, in order to detect that binding event a secondary label is introduced, such as a fluorescently labeled antibody or antigen. Such systems are usually not coupled with preparative chromatography systems as the target analyte is chemically altered by the fluorescent tag. As in the enzyme reactor detectors, these measurements are not made on a continuous basis. One group has reported some success in applying a fluorescence immunoassay in a continuous format (3). In that report, the eluant from the column was mixed with a fluorescently-labeled antibody to the analyte of interest. Excess labeled antibody was removed with a rapid affinity separation on an immobilized-analyte column so that labeled-antibody bound to analyte was measured. More recently, this same group of investigators used labeled antigens or ligands instead of labeled antibodies (3). Excess antibody was allowed to react with the analyte in the sample. To determine the amount of unbound antibody, labeled antigen was introduced and free labeled antigen was separated from the bound fraction by passing the solution through a C_{18}-silica restricted access support and then through a fluorescence detector. A large signal indicated a small concentration of analyte.
These approaches, while effective, require that one or two reagents be added to the column eluant, and also require additional columns and/or reaction chambers to remove the free tagged species before the measurement can be made.

To circumvent the limitations of these techniques, we have used a Surface Transverse Wave (STW) device as an on-line detector for liquid chromatography. The device can be derivatized to bind the target compound of interest. The sensing chemistry can consist of an antibody, antigen, Protein A or Protein G or any receptor or ligand capable of binding its complementary partner. One of the key features of the STW device is its ability to make a continuous measurement. In this report, we present the STW on-line detection system as an example of a "reagentless" detector which can make continuous measurements without the need for other separation steps.

Experimental

Chromatography System: A Hewlett-Packard 1050 Liquid Chromatography System (HP1050) with a Quaternary Pump was used for separations. The separations were monitored with a Hewlett-Packard 1050 Variable Wavelength Detector (280 nm) and an autosampler was used to introduce samples into the column. For this application a TSK-Gel column (SP-5PW 07161, TosoHaas, Stuggart, Germany) with dimensions 7.5 cm X 7.5 mm was used. Data were acquired with the Hewlett-Packard ChemStation software which controls the HP1050 Liquid Chromatograph.

Surface Acoustic Wave (STW) Devices and electronics: Specific details of the STW devices and the electronic and physical properties of these devices can be found in previous publications (7,8). Briefly, the STW device is a piezoelectric quartz crystal (0.5 mm ST-cut quartz) upon which interdigitated transducers (IDT's) have been microfabricated to produce a transmitter and a receiver. The application of a radio frequency signal of 250 MHz to the transmitter results in the launching of an acoustic wave which is received by the receiver. A layer of SiO_2 (500 A) protects the IDT's and the grating from corrosion and provides a point of attachment for performing immobilization chemistries. The electronic portion of the system consists of a radio frequency interferometer which converts changes in phase into changes in voltage. The sensing device is located in one arm of the interferometer and the reference device is located in the other. In this configuration, the output voltage is proportional to the difference in phase between the two devices. An A/D Converter (HP 35900, Hewlett-Packard, Palo Alto, CA) was used to read the signal into the HP ChemStation software. Figure 1 illustrates the HPLC-LC detection configuration, which includes the chromatography system and the STW system.

Reagents and Conditions: Sodium acetate buffer (20 mM, pH 6.0, J.T. Baker Chemical Co., Phillipsburg, N.J.), was prepared as solvent (A). Solvent (B) was solvent (A) plus 1 M NaCl (J.T. Baker). The gradient was run from 100% (A) to 100% (B) over a period of 30 minutes. The sample size was 45 ul in solvent (A) and the flow rate was 1 ml/min. Human IgG (HIgG) was purchased from Cappel Laboratories (Westchester, PA) and Fetal Bovine Serum (FBS) from Sigma Chemical Co. (St. Louis, MO). Samples included Human IgG at 1 mg/ml in solvent (A) and 100 ug/ml of HIgG in solvent (A) plus 10% FBS. For chemical immobilization of the STW devices, the following reagents were used: 3-

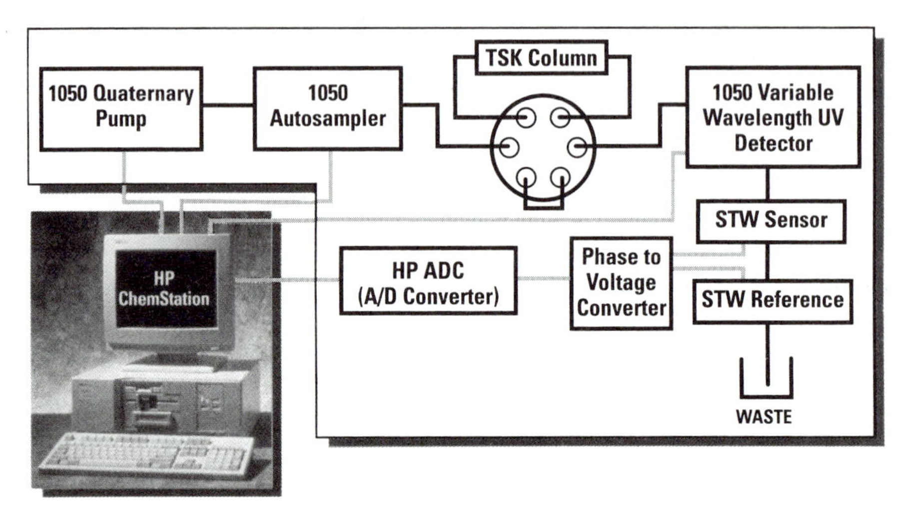

Figure 1. Schematic of the HPLC sensor configuration.

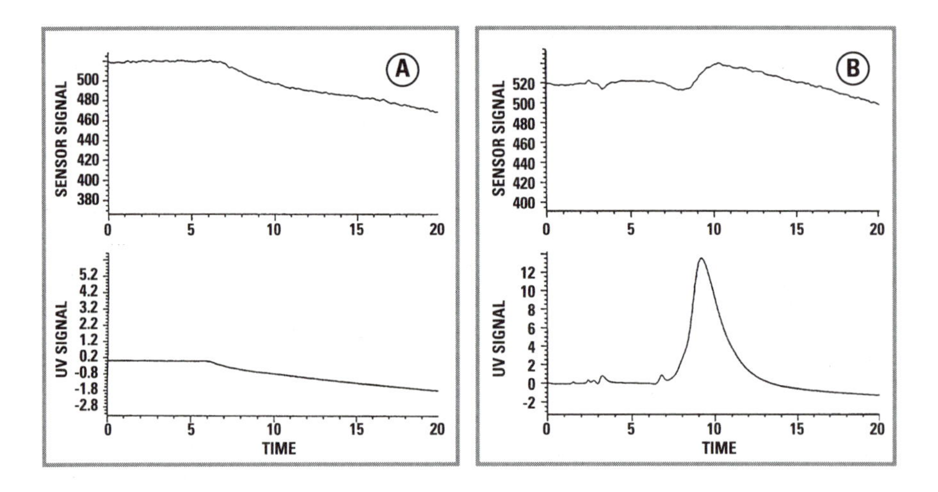

Figure 2. Comparison of blank and Human IgG (1 mg/ml) chromatograms. Panel (A) represents the sensor signal and UV signal for a blank injection. Panel (B) represents the sensor signal and UV signal for an injection of Human IgG.

glycidoxypropyltrimethoxysilane (GOPS) (Huls America, Piscataway, N.J.), 2-propanol (J.T. Baker), triethylamine (Aldrich Chemical Co., Milwaukee,), sodium periodate (Aldrich), glacial acetic acid (Sigma), borate-buffered saline (BBS) (Sigma), cyanoborohydride (Sigma), avidin D (Vector Laboratories, Burlingame, CA) and biotinylated Protein A (Pierce Chemical Co., Rockford, IL) and phosphate-buffered saline (PBS, Sigma).

Chemical Immobilization Procedure: This procedure involved coating the device with an organosilane layer, oxidizing the diol groups on the layer and incubating avidin D with the oxidized silanized device. Biotinylated Protein A was then incubated with the avidin derivatized device to produce a Protein A immobilized STW device. The specific details are as follows: A 10% solution of GOPS was made in 2-propanol containing 10% water and 4% acetic acid. The pH was adjusted to 3.0 with 1M HCl. This solution was allowed to hydrolyze for 1 hour before 0.25 ml triethylamine was added. The devices were incubated with 100 ul of the modified silane for 1 hour at room temperature. The devices were then washed with distilled water and placed in an oven at 110^0 C for 10 min to allow the silane to cure. Silanization with GOPS resulted in the formation of diol groups on the derivatized surface. To oxidize the diol groups to form aldehyde groups, a solution of 0.1% sodium periodate was made in acetic acid containing 20% water. The GOPS-derivatized devices were incubated for 30 minutes in the sodium periodate solution and then washed with borate-buffered saline pH 8.5. Avidin D (100 ug/ml in BBS, pH 8.5) was then incubated with the devices overnight at 4^0C. The next day, a solution of 0.1M $NaCNBH_4$ in 50 mM phosphate buffer, pH 6.0 was prepared and added to the avidin D solution already on the device for a final concentration of 0.01M $NaCNBH_4$. The reaction was allowed to proceed for 30 minutes before the devices were rinsed. At this point, the avidin-derivatized devices could be prepared with biotinylated Protein A. Biotinylated Protein A was reconstituted in distilled water according to the manufacturer's directions and was used at a concentration of 2 ug/ml. The incubation proceeded for 4 hours at room temperature, after which the devices were washed in PBS and ready for use. The reference device used in the experiments had avidin D only immobilized on the surface.

Results and Discussion

A sample containing 1 mg/ml HIgG in sodium acetate buffer pH 6.0, and a second sample containing the buffer alone (blank) was loaded onto the column using the autosampler. The flow rate was 1 ml/min. Figure 2 shows a comparison of the chromatograms of the blank and the HIgG sample in two panels. In both panels, the upper trace is the signal from the STW devices and the lower trace is the UV signal from the variable wavelength detector. In Panel A the sensor signal begins to drop at 7 minutes due to the effects of the gradient conditions. The UV trace also shows a slight decrease in signal due to the changes in the refractive index of the mobile phase. In Panel B, Human IgG is detected by the sensor at 8.5 minutes when the signal begins to rise. This corresponds to the strong UV signal seen in the lower panel. Due to the effects of the gradient alone on the sensor signal we chose to

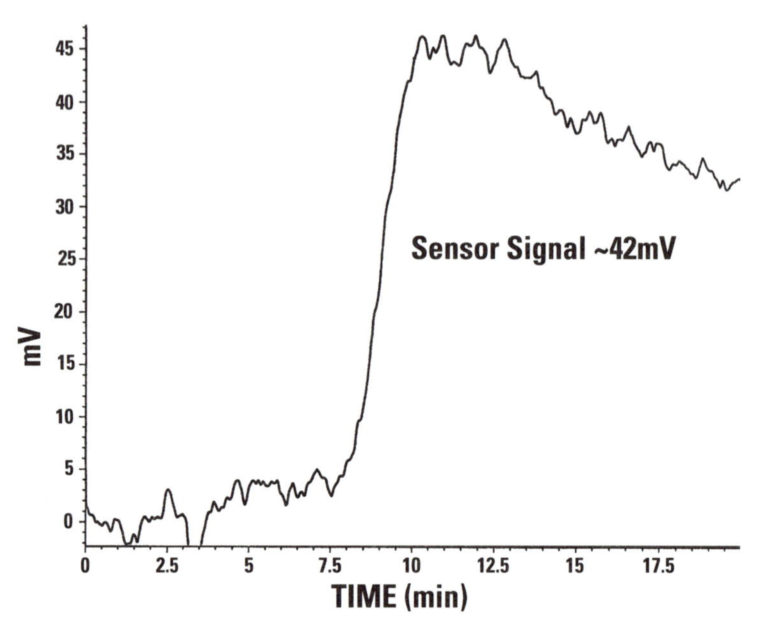

Figure 3. Sensorgram of Human IgG (1 mg/ml) minus that of the blank. The sensor signal is about 42 mV.

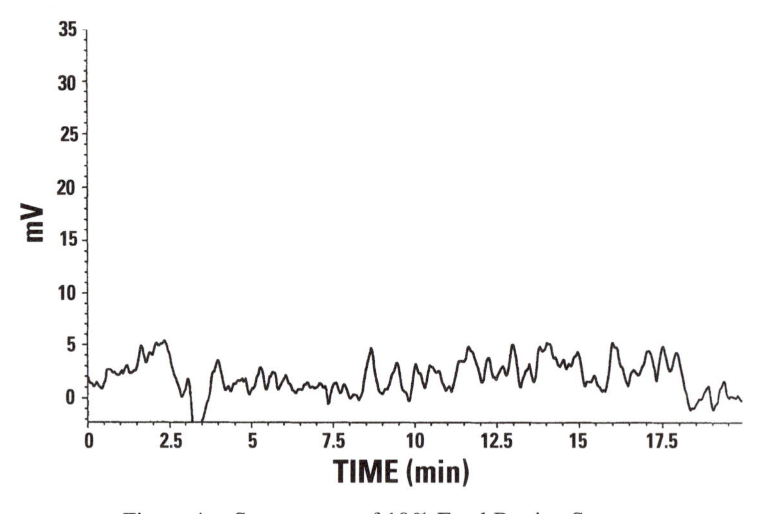

Figure 4. Sensorgram of 10% Fetal Bovine Serum.

subtract the blank chromatogram from the HIgG chromatogram to produce what we are calling a "sensorgram". This result is seen in Figure 3 which shows that 1 mg/ml of HIgG produces a phase change of approximately 42 mV. In our experiments, we refreshed the sensor by taking the column off-line using a switching valve and introduced a sample of 10 mM HCl for approximately one minute. In another experiment we ran five repeated samples of HIgG at 500 ug/ml without refreshing the sensor and found that the phase response of the last exposure was decreased about 50% from the initial exposure (data not shown). In normal chromatographic runs, it is expected that the sensors would have to be replaced after a large number of positive exposures. The replacement rate would depend on the concentration of analyte flowing across the sensors, the rate of flow and the conditions of the gradient. As was noted, there is a change in the sensor response due to the gradient alone. Although we used a reference device, the reference did not completely account for changes in ionic strength. In the case of HIgG we subtracted the blank from the sample run. However, during methods optimization, the gradient is expected to change from run to run and this could present a possible technical limitation.

To simulate potential interferences and sources of non-specific binding, we ran samples of HIgG at 100 ug/ml in 10% FBS and a blank containing 10% FBS alone. The results of the sensorgram, which represents the FBS chromatogram minus the chromatogram of a blank (buffer only), is seen in Figure 4. Note that there are no significant increases in phase and that there is negligible binding to the Protein A derivatized sensor. When HIgG (100 ug/ml) containing 10% FBS was loaded onto the column, the sensorgram of Figure 5 was produced. The sensorgram represents the difference between a blank containing 10% FBS chromatogram and the HIgG in 10% FBS chromatogram. The sensor signal was about 14 mV and with a S/N ratio of 5, is near the lowest detection limit of the sensors for HIgG using Protein A as the ligand binding protein.

As mentioned earlier, when the target compound of interest is present, the sensor generates a signal that is the integral of the analyte concentration over time. By taking the derivative of the signal, one can generate a signal that is related to the amount of analyte in the detector at any instant. This "derivative" signal resembles the signal produced by most chromatographic detectors in that it contains peaks rather than steps. An example of this is illustrated in Figure 6. This chromatogram has the UV trace for HIgG in 10% FBS (represented by the dashed lines) and the first derivative of the signal of HIgG in 10% FBS measured by the sensor (represented by the solid line). The UV trace shows several peaks corresponding to the various constituents of FBS. The HIgG peak in the UV trace is identifiable by its coincidence with the peak in the first derivative of the sensor signal. Note that the sensor peak is centered at 8.8 minutes whereas the UV peak is centered at 9.2 minutes. This time shift in the two signals is due to an artifact of the derivative calculation. However, this chromatogram illustrates how a dual detector system can identify the specific peak of interest during a chromatographic run without labels or secondary reagents.

One of the main advantages of the acoustic wave detector is the ability to make a continous measurement thereby eliminating the need to collect fractions. Other "continuous" systems are more complex, such as the continuous fluorescence

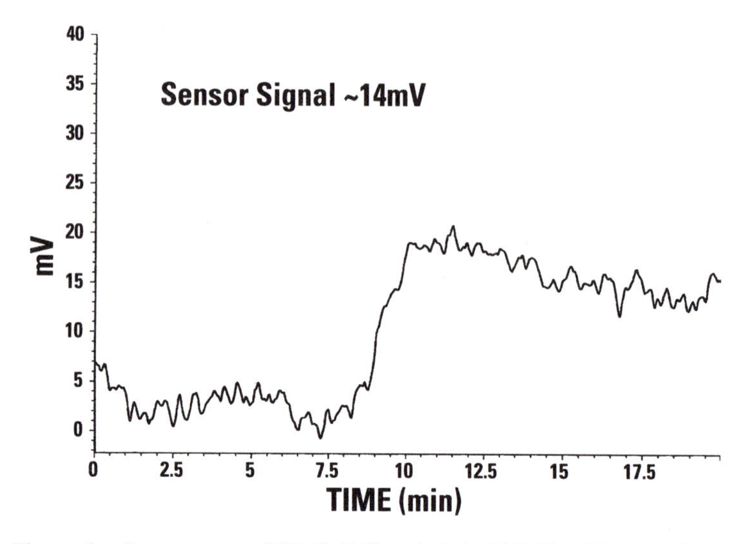

Figure 5. Sensorgram of HIgG (100 μg/ml) in 10% Fetal Bovine Serum.

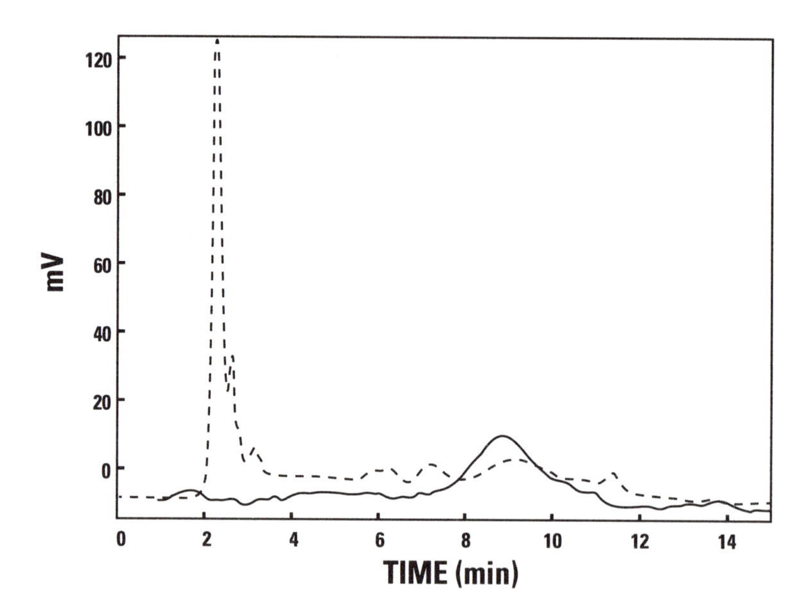

Figure 6. Chromatogram of Human IgG (100 μg/ml) in 10% Fetal Bovine Serum. The UV signal represented by the dashed lines is overlaid onto the first derivative of the sensor signal, represented by the solid line. Note the correspondence of the two signals at 9 minutes.

immunoassay reported by Oosterkamp et al., (5). In that system, the separated sample flows through two reactor chambers before being measured by the detector. In addition, reagents (both labeled and unlabeled) have to be introduced via separate pumps and then the free and bound labels need to be separated by another chromatographic column before being detected. Although the optimum reaction time for this process is between 1-2 min, the degree of complexity is high and the measurement is indirect. Our detector is in-line with the column and the eluant passes only through a UV detector and to the specific sensor. The measurement is direct and occurs in real time and no secondary reagents are needed to make the measurement. Other "reagentless" detectors include the Pharmacia BIAcore instrument which is based on surface plasmon resonance and measures the change in refractive index at or near a metal surface. The use of this instrument as a detector for LC was investigated by Nice et al. (9). Although the BIAcore system was capable of making measurements at the nanogram level, aliquots of fractions from the chromatographic run had to be collected and loaded onto an autosampler. The measurement was not continuous although it was sensitive and produced a profile of the peak fractions.

Although the STW-LC detection system has the advantage of being capable of making a continuous measurement on-line, there are some technical limitations which need to be addressed. As mentioned earlier, there was a notable change in the signal response over the chromatographic run due just to the gradient conditions alone. The reference device was unable to compensate fully for the changes in ionic strength. Good quality signals require that a blank be subtracted from each sample run. Since gradient conditions change often during methods development, and a new blank would be needed for each set of gradient conditions, the sample throughput would be low. It may be possible to modify the STW readout electronics so the signal is less responsive to changes in fluid conductivity. Another limitation is that the response of the devices decreases with each run. This may not be a problem when identification of the unknown peak is all that is needed. However, quantitation may not be possible if the response is gradually decreasing with time, unless the decrease can be predicted accurately. Finally, the flow rate in these experiments was 1 ml/min. Although higher flow rates have been successfully demonstrated on this system, higher flow rates cause an increase in the noise level of the detection system and cause a decrease in detection sensitivity, although the effect of increased flow rate on sensor performance has not been thoroughly examined. Another possible limitation is that the target analyte may elute under gradient conditions which inhibit its binding to the sensor. In this case, the target analyte cannot be detected by the sensor but can be detected by the UV-VIS spectrophotometer. Changes in ionic strength or pH could be offset by adding a diluent, but this would compromise the sensitivity of the system.

Conclusions

The feasibility of using the STW sensor as an LC detector for the separation of HIgG in buffer and 10% FBS has been demonstrated. FBS appears not to interfere with the specific detection of HIgG by the sensor. The current detection limit for HIgG is 100 ug/ml. While the system remains a research tool at the present time, the key features of the acoustic wave detector are: it is a continuous measurement in real time which does not use labels or derivatized samples, it can detect specific peaks corresponding to the target analyte during a chromatographic run, and it is quantitative. Although the system has several limitations, it has many clear advantages over the current method of fraction collection and assay.

Literature Cited

1. Galensa, G. *Food Sci. Technol.* **1994**, *60*, 191-217.
2. Marko-Varga, G.; Johansson, K.; Gorton, L. *J. Chrom.* **1994**, *660*, 153-167.
3. Oosterkamp, A.J.; Irth, H.; Tjaden, U.R.; van der Greef, J. *Anal. Chem.* **1994**, *66*, 4295-4301.
4. Hippe, H.; Stadler, H. In "*Biosensors Applications in Medicine, Environmental Protection and Process Control*"; Schmid, R.D.; Scheller, F., Eds.; GBF Monographs. VCH Publishers: New York, N.Y., 1989, Vol. 13; 289-292.
5. Oosterkamp, A.J.; Irth, H.; Beth, M.; Unger, K. K.; Tjaden, U.R.; van de Greef, J. *J. Chrom.* **1994**, *653*, 55-61.
6. Bowers, L.D. *Chromatogr. Sci.*, **1986**, *34*, 195-225.
7. Baer, R.L.; Flory, C.A.; Tom-Moy, M.; Solomon, D.S. *Proc. IEEE Ulrason. Symp.* **1992**, 293-298.
8. Tom-Moy, M.; Baer, R.L.; Doherty, T.P.; Solomon, D.S. *Anal. Chem.* **1995** (in press).
9. Nice, E.; Lackmann, M.; Smyth, F.; Fabri, L.; Burgess, A.W. *J. Chrom.* **1994**, *660*, 169-185.

RECEIVED July 20, 1995

Chapter 3

Immunosensors for Detection of Chemical Mixtures

Antibody Affinities, Selectivities, and Cloning

Mohyee Eldefrawi[1], Amira Eldefrawi[1], Jeremy Wright[2], Peter Emanuel[3], James Valdes[3], and Kim R. Rogers[4]

[1]Department of Pharmacology and Experimental Therapeutics, University of Maryland School of Medicine, Baltimore, MD 21201
[2]Department of Pharmaceutical Sciences, University of Maryland School of Pharmacy, Baltimore, MD 21201
[3]U.S. Army Edgewood Research, Development, and Engineering Center, Aberdeen Proving Ground, MD 21010
[4]National Exposure Research Laboratory, Characterization Research Division, U.S. Environmental Protection Agency, P.O. Box 93478, Las Vegas, NV 89193–3478

An important objective in using biosensors is to accurately and rapidly detect and quantitate analytes of interest. When using antibodies as biological recognition elements, the specificity, reversibility, and to some extent the detection limits of a biosensor are determined by the characteristics of the antibodies used. Similar to the high specificity afforded by subpopulations of polyclonal antibodies directed toward different haptenic determinants on the same antigen, the use of combinatorial cDNA libraries to clone Fab fragments may provide the ability to tailor this type of specificity.

Immunoassays are based on the high affinity recognition capabilities of antibodies (Abs) for their antigens (Ags). Radioimmunoassay (RIA) requires high affinity binding of Ab to Ag so that the complex may be isolated from a chemical mixture in solution, by means of the high affinity binding of a secondary Ab (Van Heyningen et al., 1983). Similarly, the enzyme linked immunosorbent assays (ELISAs) also require high affinity binding of Ab to the Ag. This is required so that Ab-Ag complexes immobilized in the wells of a microtiter plate can be separated from other components in the sample mixture (Douillard and Hoffman, 1983; Kaufman and Clower, 1991). RIA and ELISA have been applied to the detection of small molecular weight compounds as well as larger protein Ags (Shiveley et al., 1983 Yao and Mahoney, 1984; Verbey and DePace, 1989). Immunosensors using Abs, that are immobilized on different transducers, have utilized both polyclonal Abs (Anis et al., 1993) and monoclonal Abs (mAbs) (Oroszlan et al., 1993; Devine et al., 1995).

0097–6156/95/0613–0019$12.00/0

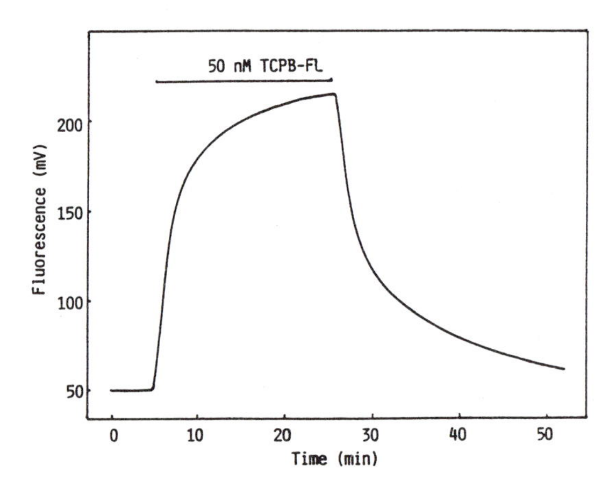

Fig. 1 - The time course of association of TCPB-FL to and its dissociation from the Ab-coated fibers. Bar indicates the time that TCPB-FL (50 nM) was present in the PBS flow buffer (reproduced with permission from reference 27, Copyright 1995, American Chemical Society).

The affinity of an Ab for an Ag is a measure of the stability of the Ab-Ag complex. The binding of Ag to Ab proceeds according to the Gibbs standard free energy change ($\Delta G° = \Delta H°-T\Delta S°$) or in terms of affinity parameters ($\Delta G° = -RT \ln KA$); where R = gas content (i.e. 1.99 cal/mol. deg); T = degrees in Kelvin; KA = equilibrium association constant; $\Delta H°$ = the enthalpy term; and $\Delta S°$ = the entropy.

Although large negative ΔG values reflect high affinity of the Ab for the Ag, immunosensor design must also take into consideration the binding kinetics. In order for a biosensor to be useful as an analytical monitoring device, it must respond to changes in analyte concentration reversibly and within a relatively short time period (i.e. 1-5 min). Because Ab-Ag binding typically requires several hours to reach steady state, biosensor assay parameters are usually adjusted to allow measurements to be made prior to steady state. Consequently, the ability to determine kinetic parameters of the biosensor system is critical to assay format and design.

Use of Immunosensors to Measure the Antibody Affinity for its Antigen

In addition to analytical tools, biosensors have also been used to measure Ab-Ag affinities. Examples include: surface plasmon resonance and fiber optic technologies (Pellequer and Van Reganmortel, 1993; Devine et al., 1995). Among the fiber optic biosensors, evanescent wave methods lend themselves to kinetics analysis. For the evanescent wave biosensor, a fluorescent analog of the analyte is typically used as the optical signal generator. This approach allows the observation (in real time) of the association and dissociation of the labeled Ag to the surface-immobilized Ab. For example, the binding of 2,4,5-trichlorophenoxybutrate (TCPB) fluorescein (FL) to anti-polychlorinated biphenyl (PCB) Abs immobilized to a quartz fiber surface has been used to determine several kinetic parameters. Upon introduction of TCPB-FL to the Ab-coated fiber, the observed fluorescence increases rapidly as TCPB-FL binds to the Ab-coated fiber and reaches a steady state in <20 min (Fig. 1). The observed fluorescence reflects the real-time binding of TCPB-FL to the immobilized Abs (the slope of the rising portion of the curve is defined as fluorescence vs time). Since the amount of bound TCPB-FL is very small (<0.1%) compared to the concentration of TCPB-FL flowing past the Ab-coated fiber, the fluorescent-tagged probe concentration is not depleted as a result of binding and the pseudo first order rate constant the apparent association constant (k_{app}) can be calculated from the slope of the rising portion of the curve. Removal of TCPB-FL from the flow buffer, after steady state fluorescence has been reached, results in an immediate decrease in fluorescence. The dissociation rate constant (k_{-1}) can be determined by measuring the halftime ($t_{1/2}$) of dissociation of TCPB-FL from the fiber. The $t_{1/2}$ is determined from a semi-logarithmic plot of fluorescence (mV/min), after removal of TCPB-FL from the buffer, and k_{-1} is calculated using the formula $k_{-1} = 0.639/t_{1/2}$. The k_{app} value is obtained from the linear plot of ln [FLss/(FLss-FL$_t$)] vs time, where FLss is steady state fluorescence and FL_t is fluorescence recorded during steady state binding. The association rate constant (k_{+1}) can then be calculated from the formula: $k_{+1}=(k_{app}-k_{-1})$ / [TCPB-FL]. The equilibrium dissociation constant K_D can then be calculated from the formula $K_D=k_{-1}/ k_{+1}$. K_D is the reciprocal of K_A.

Specificity of the Immunosensor

It is generally considered that Abs bind their Ags with high specificities. However, both polyclonal and monoclonal Abs in general will bind (with varying degrees of affinity) to compounds structurally related to the antigen. In particular, for small molecular weight haptens, polyclonal antibodies may be directed toward the protein used as the carrier as well as to the hapten and chemical structure used to link them. This is an important consideration for certain immunosensor assay formats.

Demonstration of immunospecificity and optimization of the assay for development of evanescent wave biosensors entails several steps. The choice of a fluorescently labeled Ag typically requires the evaluation of the binding characteristics of several tracers which are structurally related to the antigen. This allows the optimal choice of a fluorescent tracer. In the case of evanescent biosensors, the fluorescently labeled antigen tracer must bind to the antibody with sufficient affinity so that a relatively low concentration (e.g., 1 to 100 nM) will yield a reproducible signal, yet show a relatively low affinity as compared to the target analyte to be measured.

In another set of experiments, the specificity with which the fluorescent analyte binds to non-specific Abs and unrelated proteins (e.g. bovine serum albumin, casein) or polymers (e.g. collagen, cellulose) is measured. This non-specific binding of the Ag tracer is then minimized and/or "adjusted for" in the binding measurements.

The data in Figure 2 (left) demonstrates that the mAb directed toward benzoylecgonine (BE) binds to fluorescein-labeled BE (BE-FL) with an affinity high enough that a relatively low concentration (i.e. 10 nM) yields a substantial fluorescence signal (over the background noise). At this BE-FL concentration, relatively little non-specific binding is observed to human IgG, casein, or the quartz fiber alone. In contrast to the low concentration of BE-FL fluorescent tracer required to observe a signal using the anti-BE mAb, a polyclonal anti-PCB Ab system required a higher concentration (50 nM) of the fluorescent antigen probe TCPB-FL. The higher probe concentration required for this system contributed to a higher level of non-specific (i.e. Ab-independent) signal (Figure 2, right).

Sensitivity and Detection Limits

The sensitivity of the immunosensor is determined primarily by the affinities that the Ab has for the fluorescent tracer and the analyte. The higher the affinity of the Ab for the Ag, the higher the sensitivity and increased ability to detect lower concentrations. The detection limit is defined as the lowest concentration of the analyte that results in significant inhibition of binding of the fluorescent probe (i.e. two standard deviations below the mean value of control). For Abs, higher affinities generally reflect a slower rate of dissociation of the analyte from the Ab and better ability to compete with the fluorescent tracer for binding to the immobilized Abs. In the case of Abs used in the development of immunosensors for monitoring purposes, Ab affinities must be considered in recognizing the compromise between the detection limit and reversibility. With the use of appropriate fluorescence probes and operating formats, however, both reasonable reversibility and low detection limits can be achieved.

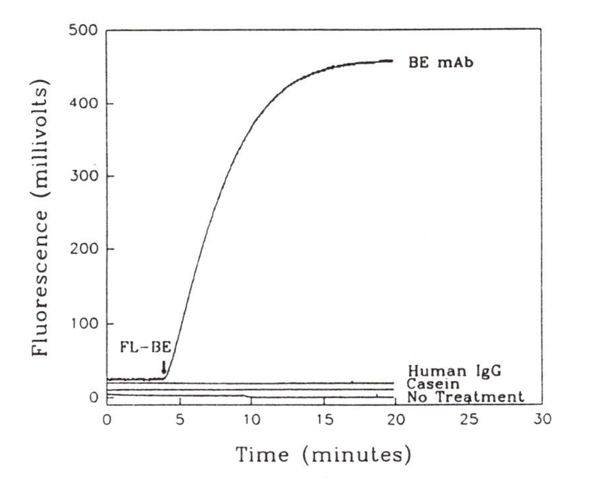

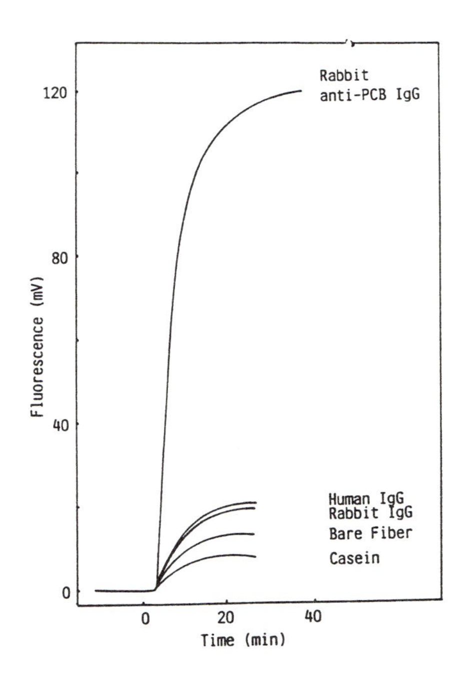

Fig. 2 - The specificity of binding of the fluorescent probe to the Ab-coated optic fibers of the biosensor. (left) The fluorescent signal transmitted by quartz fibers coated with mAbs raised against BE, compared to that of untreated fibers and fibers coated with human IgG or casein. The flow buffer contained 10 nM FL-BE. The lines are offset from each other by 10 mV. Reproduced with permission from reference 4 Copyright 1995 American Chemical Society. (right) The fluorescence transmitted by TCPB-FL, bound to the fiber coated with rabbit anti-PCB IgGs, control rabbit IgG, human IgG or casein compared to a bare fiber. It is clear that the non-specific TCPB-FL binding to the quartz fibers, non-target IgG can be significant and must be corrected for. Reproduced with permission from references 4 and 27, Copyright 1995, American Chemical Society.

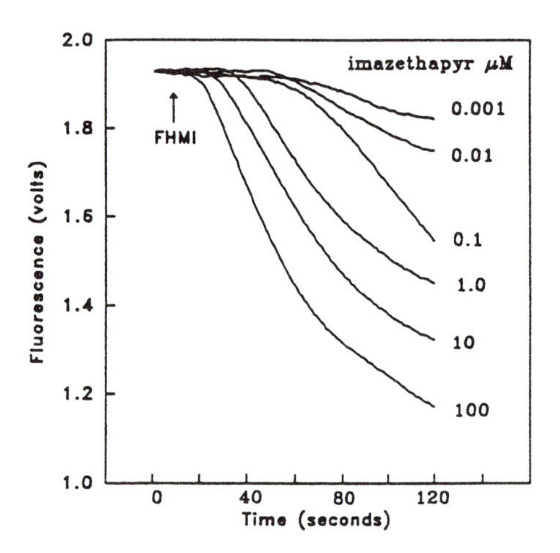

Fig. 3 - Displacement of the fluorescent probe fluorescein hydrozino methylene imazepyr (FHMI) by imazethapyr. The Ab-coated fiber was perfused with the flow buffer (PBS/casein, 25 nM FHMI until fluorescence reached a steady state (<10 min). The FHMI was removed and the fiber was perfused with PBS/casein containing imazethapyr for 2 min then the flow buffer was reintroduced. Fluorescence recovered back to the steady state level. This process was repeated 5 more times using increasing concentrations of imazethapyr. In other words a single fiber was used to generate all the data. Reproduced with permission from reference 2, Copyright 1993 American Chemical Society.

For monitoring purposes, biosensors should be easy to regenerate after making a measurement, so that the same biosensor can be used repeatedly. In a flow injection mode, the best scenario is a few minutes of wash with analyte free buffer between measurements. An optimal compromise between reversibility and sensitivity was observed with the fiber optic immunosensor for detection of imazethapyr herbicide (Anis et al., 1993). Partially purified polyclonal Abs, produced against an imazethapyr derivative conjugated to bovine serum albumin, were used to coat the quartz fibers. A fluorescent conjugate of the same imazethapyr derivative bound to the Ab-coated fiber with relatively high affinity and this binding was displaced easily with imazethapyr (Fig. 3). Removal of the analyte (i.e. imazethapyr) from the flow buffer resulted in rapid biosensor regeneration. Similar findings were reported recently using a fiber optic immunosensor for cocaine (Fig. 4). In the latter case, FL-benzoylecgonine (BE), a metabolite of cocaine, was used as the fluorescent probe and a mAb was used as the recognition element. Its detection limit for cocaine was 5 ng/ml (Devine et al., 1995).

Cross Reactivity of Other Constituents in a Chemical Mixture

Another important feature of an immunosensor is the Ab selectivity for the analyte of interest relative to structurally related chemicals, that may be present in the samples to be tested. Depending on the potential application for an immunosensor, various ranges of target analyte specificity may be advantageous. For example, in detection of drugs of abuse, the application may require the detection of a certain metabolite or closely related group of metabolites. This may also be the case for detection of a particular pesticide or closely related group of environmental break-down products. In contrast to detection of clinical analytes, for environmental pollutants such as PCBs, a relatively high degree of cross-reactivity among the 209 congeners would be of considerable value.

A useful measure of cross-reactivity is the equilibrium constant of inhibition Ki (Budd, 1981). The Ki value is calculated using the Cheng-Prusoff (1973) equation: $Ki = IC_{50} /(1+[I]/K_D)$. The IC_{50} value is the analyte concentration which produces 50% reduction in the observed fluorescence signal (i.e. reduction in binding of the analyte-tracer) and is derived from the analyte concentration-response function. Ki values of several coca alkaloids and cocaine metabolites were used to determine cross-reactivities using a fiber optic cocaine biosensor, which used a mAb as a recognition element and BE-FL as the fluorescent tracer (Table 1).

In the application of this biosensor to the detection of urinary metabolites of cocaine, there are certain advantages and limitations compared to standard chromatographic analysis. Although speed, convenience, and sensitivity are some of the advantages, the cross-reactivities of several cocaine metabolites and derivatives must be considered. It is evident that this mAb, if used in commercial kits for urine analysis, could detect cocaine at a 6-fold lower concentration than it would BE, against which the mAb was raised. The urine of a cocaine user contains primarily the major metabolite BE and very little cocaine (Devine et al., 1995). In addition, if the user ingests alcohol with cocaine the urine would also contain cocaethylene. In this situation, the high affinity that this mAb has for cocaethylene would result in as much as a 6-fold overestimate of the amount of BE in urine and consequently of cocaine administered by the user. The limited

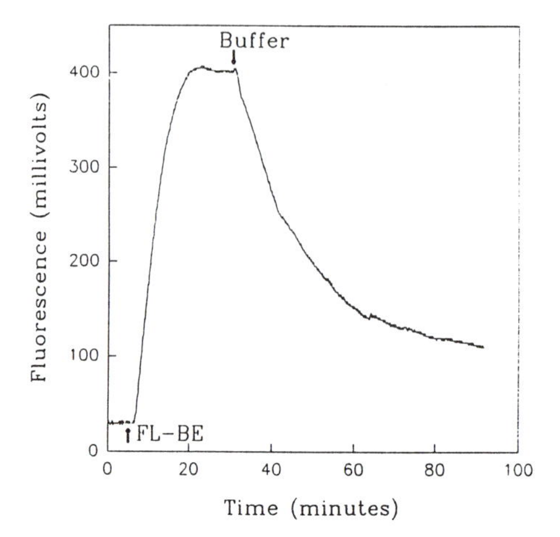

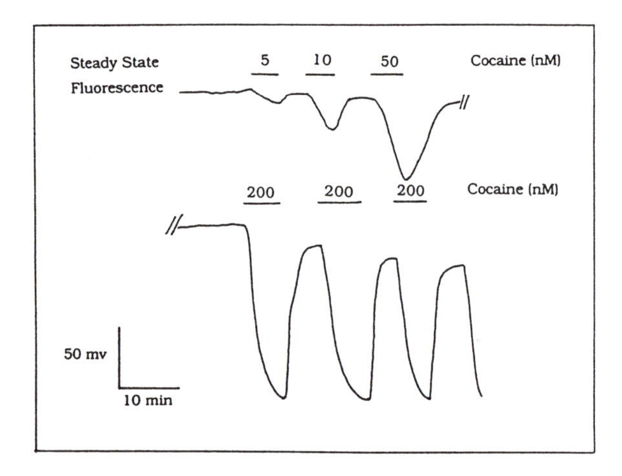

Fig. 4 - An optic fiber biosensor for cocaine using a mAb against benzoylecgonine as the biological sensing element. (left) The time course of binding of FL-BE to the mAb-coated fiber expressed by the fluorescent signal transmitted via the fiber. After reaching steady state, FL-BE was withdrawn from the flow buffer (indicated by the arrow). The bound FL-BE dissociated and fluorescence decreased exponentially. (right) Reusability of the biosensor for multiple assays of cocaine introduced into the flow buffer after steady-state fluorescence (200 mV) was achieved. Cocaine at the indicated concentrations was added to the flow buffer for only the time intervals indicated by the bars. The downward deflection resulted from displacement of FL-BE by cocaine, but upon removal of cocaine from the flow buffer FL-BE displaced the bound cocaine. Reproduced with permission from reference 4, Copyright 1995 American Chemical Society.

data suggest that structural features that are necessary for high-affinity recognition of cocaine metabolites by this mAb are a benzoate (absent in ecgonine and methylecgonine) and 3β configuration, but lacks the carboxymethyl moiety of BE. The biosensor also has high affinity for tropacocaine, which has the 3 benzoate moiety, but not the carboxymethyl moiety of cocaine at C-2 or the carboxyamide moiety of BE. This biosensor also has excellent stereospecificity with 833-fold lower affinity for pseudococaine than for its stereoisomer cocaine. Coca leaf extracts contain several coca alkaloids, with methylecgonine and tropacocaine representing 29-57% and 0.16-3.8% of cocaine concentration, respectively (Johnson and Emche, 1994; Moore and Casale, 1994). If this same biosensor is used for detection and quantitation of these alkaloids, it is expected to hardly detect methylecogonine and dramatically overestimate tropacocaine.

Table 1. Detection Limits and Equilibrium Constants of Inhibition of Cocaine Metabolites and Structurally Related Compounds Using Inhibition of Binding of FL-BE to mAb-Coated Fibers

Chemical	Detection limit (ng/ml)	K_i[a] (nM)	Cross-reactivity[b]
Metabolites			
Cocaine	5	30	600
Benzoylecgonine	30	180	100
Cocaethylene	5	46	391
Norcocaine	29	810	22
Methylecgonine	2,000	100,000	0.18
Ecgonine	4,600	300,000	0.06
Related compounds			
Tropacocaine	10	62	290
Pseudococaine	1,500	25,000	0.72
Atropine	2,900	28,000	0.64
Sodium benzoate	>9,999	>900,000	0.02

[a]For each experiment, six different analyte concentrations were tested on six different fibers, and then the K_i value was calculated from two such experiments for each of the above compounds (difference ranged from 50 to 10%), except for cocaine and BE whose K_i values were calculated from four and three separate experiments, respectively. Their mean Ki+ SE values are 30± 3 and 180 ± 11 nM, respectively.

[b] Cross-reactivity was calculated as the ratio of the K_i value for displacing FL-BE by BE to the K_i value for displacing FL-BE by the test drug, multiplied by 100. Reproduced with permission from reference 4, Copyright 1995 American Chemical Society.

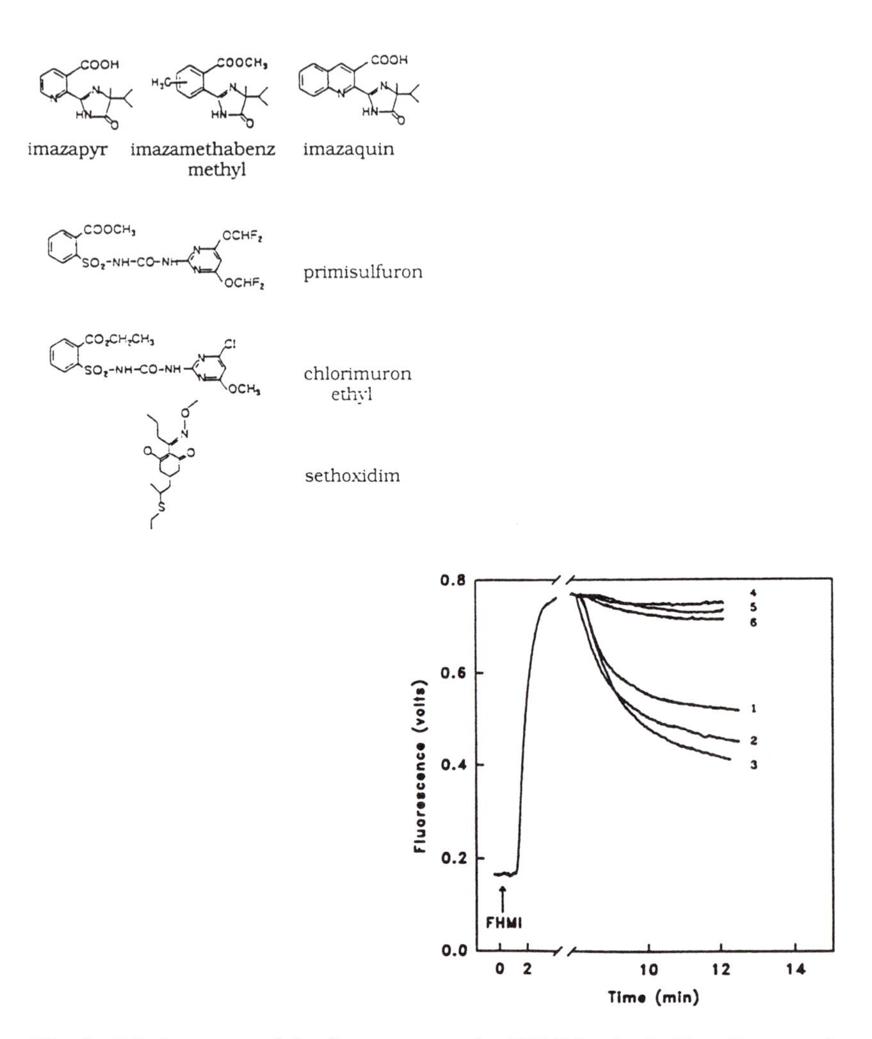

Fig. 5 - Displacement of the fluorescent probe FHMI by the imidazolinone and non-imidazolinone compounds. (left) Chemical structures of three imidazolinone class compounds and three non-imidazoline agrochemicals used to demonstrate the specificity of the fiber optic sensor. (right) Reduction of steady-state fluorescence by the various analytes. A 25 nM FHMI in casein/PBS solution was perfused to reach a steady state before a solution of 1 μM compound was added to the FHMI, casein/PBS. Plot 1, imzamethabenz methyl; 2, imazapyr; 3 imzaaquin; 4, chlorimuron ethyl; 5, primisulfuron; 6, sethoxydim. Reproduced with permission from reference 2, Copyright 1993 American Chemical Society.

Another example of a cross reactivity profile for a fiber optic immunosensor that has been observed is for the herbicide imazethapyr, using a polyclonal Ab. This biosensor detected several imidazolinone herbicides (at low concentrations), but had low cross reactivities for non-imidazolinone compounds (Fig. 5). It is thus considered to be a class-specific biosensor that will detect but not distinguish well amongst imidazolinone compounds. In a recent study polyclonal Abs raised against a particular PCB congener were used in an immunosensor to detect PCBs and other chlorinated hydrocarbons (Zhao et al., 1995). This immunosensor also behaved as a class specific biosensor because of its high cross reactivity with numerous PCBs, but showed remarkably lower responses towards other chlorinated hydrocarbons that may be present in environmental samples. With respect to immunosensor selectivity, the use of Ab cloning and engineering may provide a wide variety of Abs with a range of selectivities which could potentially be used to tailor the specificity of the biosensor for a variety of particular applications.

Cloning of Antibodies

Antigens present multiple epitopes to the immune cells when injected in an animal resulting in the production of many Abs, each coming from a different immune cell. Thus, the serum fraction of the blood of an immunized animal contains polyclonal Abs, that exhibit differing affinities for the Ag and other chemicals in a mixture. Small chemicals (haptens) such as drugs, pesticides and other chemicals of environmental concern (e.g. PCB), must first be conjugated to carrier proteins, before use as antigens to produce Abs that recognize the haptens (Langone and Van Vunakis, 1983). These antigens trigger the formation of polyclonal Abs, only a few of which target the hapten molecule (i.e. the chemical of interest) and with varying affinities. The highest affinity Ab would be the most desirable to use in RIA and ELISA assays, so as to detect the lowest analyte concentration and increase its specificity. However, a lower affinity Ab may be desirable to use in immunosensors developed for monitoring purposes (Anis et al., 1993; Oroszlan et al., 1993), because lower affinity allows the Ab-Ag complex to dissociate faster for quick regeneration of the sensor. Also, since the array of Abs produced varies from animal to animal and even between bleedings of the same animal, in order to insure a continuous supply of the same Abs, mAbs are typically better suited for development of immuno-assays than polyclonal sera.

In 1975, Kohler and Milstein described a method for making cell lines that secrete a single species of Ab (i.e. mAb) with the desired specificity to a particular antigen (Milstein, 1980). The technique known as hybridoma technology describes fusion of β lymphocytes with myeloma cells that do not produce immunoglobulins in presence of 35% polyethylene glycol. Fused cells are selected for the ability to grow on hypoxanthine, aminopterin, and thymidine containing medium. Cells that produce Abs (i.e. hybridomas) are identified by immune assays and are individually sub-cultured. Hybridoma technology was successfully applied to the production of a wide range of mAbs that bind to proteins, and small haptens conjugated to carrier proteins and also to Abs that even have catalytic activities; leading to many practical applications for mAbs in research and human health care. However, hybridoma cells, like most other animal cells in culture, grow relatively slowly, do not attain high cell densities and require complex and expensive growth medium. The cost of their production is an impediment to the more widespread use of mAbs in diagnostic procedures. To circumvent these problems, attempts have been made to genetically engineer bacteria, as "bioreactors", for the production of mAbs (Pluckthum, 1991; Ward, 1992).

Complementary DNA (cDNA) is synthesized from RNA isolated from mouse antibody-producing cells (B-lymphocytes) and the heavy and light chain sequences (coding for different portions of the variable, Fab, region of the Ab) are selectively amplified using the polymerase chain reaction (PCR) procedure (Fig. 6A). The amplified cDNA preparation is digested with a specific set of restriction endonucleases and randomly clone together to form antibody-gene cassettes which are then ligated into a bacteriophage lambda vector (Figure 6B). Directional cloning of the heavy and light chain genes is provided by engineered restriction sites contained within the oligonucleotide primers used for PCR amplification (Hogrefe, 1994). At this stage in the process, many different heavy and light chain sequences are cloned to create a population of Fab gene cassettes in a single "combinatorial" vector. During the lytic cycle of the bacteriophage lambda, the library of combinatorial clones is co-expressed forming assembled Fab fragments which are screened for antibody production or antigen binding activity (Amberg et al., 1993).

The primary lambda library is subjected to mass excision by co-infection with an ExAssist helper bacteriophage (Short and Sorge, 1992). The library is rescued in the form of "phagemids" which contain sequences that allow the construct to exist either as a double stranded plasmid in bacteria or to package themselves into filamentous bacteriophage. The phagemid has had critical sequences needed for the bacteriophage packaging deleted. When co-infected with VCSM13 helper virus, the phagemid directs itself to be packaged into filamentous bacteriophage. The recombinant Fab antibody gene is designed to be expressed as a fusion with a coat protein of the bacteriophage in such a way as to "display" the gene product of interest on its surface. The surface display of the Fab allows for the enrichment of high affinity clones through biopanning (Figure 6C) whereas the vast majority of phages that do not bind with high affinity are discarded (Barbas, 1991; Garrard et al., 1991; McCafferty et al., 1990). The Fab antibody genes are excised on the phagemid and used to transform *E.coli*. The Fab polypeptides generated by the *E.coli* are harvested and purified by means of an engineered histidine tag using standard IMAC columns (Hochuli et al., 1987; Hoffman et al, 1991).

The step in which light and heavy chain cDNAs are combined on one vector creates a vast array of diverse Ab genes, some of which will encode unique target-binding sites that would never have been possible to isolate with standard hybridoma procedures. The mammalian Ab repertoire has the potential of producing approximately 10^6 to 10^8 different mAbs. A phage library contains approximately this number of clones, so one combinatorial library can be expected to produce as many different mAbs (Fab molecules) as mammalian systems. In addition, once an initial combinatorial library has been constructed, it is possible to shuffle the light and heavy chains to obtain Fab molecules that recognize unusual epitopes. An even greater variation can be achieved by mutagenesis of the combinatorial library. Because millions of bacteriophage plaques can be screened in a relatively short period of time, the identification of Fab molecules with the desired specificity takes only 7 to 14 days. By contrast, months are required to screen a few hundred hybridoma cell lines.

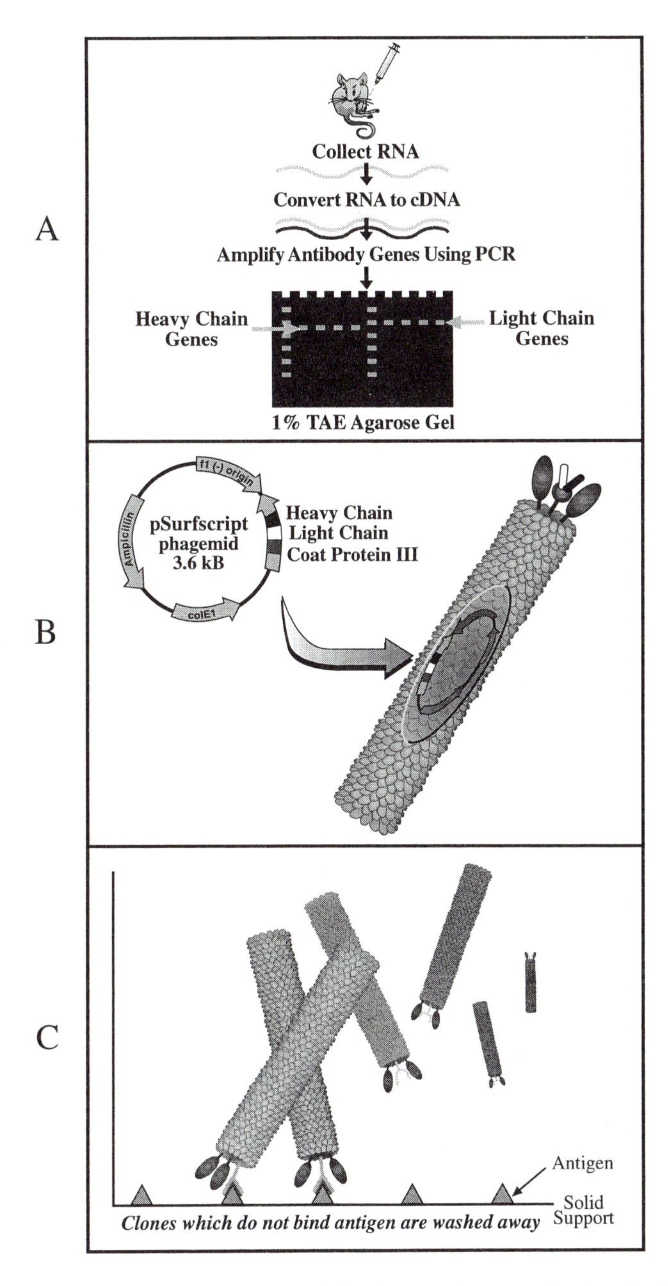

Fig. 6. Schematic representation of the three major steps in the cloning of Fab antibody fragments. A; First strand synthesis and PCR amplification. B; The pSurfscript vector excised from the primary lambda library (Amberg et al., 1993) and a cut-away view of the filamentous bacteriophage Fab protein as a fusion with the phage coat protein. C; Biopanning enrichment of positive clones.

Acknowledgements and Notice

Much of the underpinnings of the above work was funded by U.S. Army contract No. DAAM01-94-C0020. Some of the information in this writeup was funded in part by the Environmental Protection Agency under cooperative agreement No. CR820460-0, with the University of Maryland at Baltimore, NIH grant no. R01-DA08222 and USDA agreement no 58-1275-3-018.

Literature Cited

1. Amberg, J.H.; Hogrefe, H. Lovejoy, B.H.; Shopes, B.; Mullinax, R.; Sorge, J. *Strategies in Molecular Biology* 1993, 5, 2-5.
2. Anis, N.A.; Eldefrawi, M.E.; Wong, R.B. *J. Agri. Food Chem.*, 1993, 41, 843-848. Barbas, C.; Kang, A.; Lerner, R.; Benkovic, S. *Proc. Natl. Acad. Sci.* 1991, 88, 7978-7982.
3. Cheng, Y.; Prusoff, W.H. *Biochem. Pharmacol.* 1989, 182, 353-359.
4. Devine, P.J.; Anis, N.A.; Wright, J.; Kim, S.; Eldefrawi, A.T.; Eldefrawi, M.E. *Analyt. Biochem.*, 1995, 227, 216-224
5. Douillard, J.Y.; Hoffman T. *Meth. Enzymol.* 1983, 92, 168-174.
6. Garrard, L.; Yang, M.; O'Connell, M.; Kelley, R.; Henner, D. Bio/Technology 1991, 9, 1373-1377.
7. Glass, T.R.; Lackie, S.; Hirschfeld, T.B. *Appl. Opt.* 1986, 26, 2181-2187.
8. Hochuli, E; Dobeli, H.; Schacher, A. *J. Chromat.* 1987, 411, 177-184.
9. Hoffman, A.; Roeder, R. *Nucleic Acids Res.* 1991, 19.22, 6337-6338.
10. Hogrefe, H.A.B.S. *PCR Methods and Applications* 1994, S109-S122.
11. Johnson, E.L.; Emche, S.D. *Ann. Bot.* 1994, 73, 645-650.
12. Kaufman, B.M.; Clower, M. *Jr. Assoc. Off. Anal. Chem.* 1991, 74, 239-247.
13. Kohler, G.; Milstein, L., *Nature*, 1975, 256, 495-497.
14. Langone, J.J.; Van Vunakis, H. *Methods Enzymol.*, 1982, 628-640.
15. McCafferty, J; Griffiths, A.D.; Winter, G.; Chriswell, D.J. *Nature* !990, 348, 552-554.
16. Moore, J.M.; Caslae, J.F. *J. Chromatogr. A.* 1994, 674, 165-205.
17. Milstein, C. *Scientific American*, 1980, Oct. 66-74.
18. Pellequer, J.L.; Van Regenmortel, M.H.V. *J. Immunol. Methods*, 166, 133-143.
19. Oroszlan, P., Duveneck, G.L.; Ehrat, M.; Widmer, H.M. *Sens. Actuators*, 1993, 11, 301-305.
20. Pluckethum, A. *Bio/Technology*, 1991, 9:545-551.
21. Shirley, J.E.; Wagner, C.; Clark, B.R. *Meth. Enzymol.* 1986, 121, 459-472.
22. Short, J.M.; Sorge, J.A. *Methods Enzymol.* 1992, 216, 495-508.
23. van Heyningen, V.; Brock, D.J.H.; van Heyningen, S. *J. Immunol. Method*, 1983, 62, 147.
24. Verbey, K.; De Pace, A. *J. Forensic Sci* , 1989, 34, 46-52.
25. Ward, S.E. *The FASEB J.* 1992, 6, 2422-2427.
26. Yao, R.C.; Mahoney, D.F. *J. Antibiotics* 1984, 37, 1462-1468.
27. Zhao, C.Q.; Anis, N.A. Rogers, K.R.; Kline, R.H. Jr; Wright, J.; Eldefrawi, A.T.; Eldefrawi, M.E. *J. Agric. Food Chem.* 1995, (in press).

RECEIVED August 25, 1995

Chapter 4

Adaptation of a Fiber-Optic Biosensor for Use in Environmental Monitoring

Mark D. Pease, Lisa Shriver-Lake, and Frances S. Ligler

Center for Bio/Molecular Science and Engineering, Naval Research Laboratory, 4555 Overlook Avenue, S.W., Washington, DC 20375–5348

This chapter describes an evanescent wave fiber optic biosensor and its application to immunoassays for rapid detection of bacterial cells and pollutants. Whole cells of *Burkholderia cepacia* G4 5223-PR1 (G4) are of interest for their ability to degrade trichloroethylene (TCE) which is one of the most prevalent contaminants of ground water in the United States. The lower limit of detection of the G4 with this system is 10^4 - 10^5 cells/ml. In addition to TCE, the explosive trinitrotoluene (TNT) is a known contaminant of ground water. Limits of detection of TNT with this system is 10 ng/ml.

As governments and individuals become increasingly conscience of the limited and fragile supply of resources available on the earth, increasing attention is focused on preserving and regenerating our environment. Decades ago the phrase "environmental monitoring" conjured up images of testing the water down stream of industrial plants, or checking lead levels in the air. Today however, the realities of environmental monitoring are much more diverse, positive, and proactive. This chapter focuses on one technology which has been adapted to meet the needs of environmental monitoring; specifically it discusses the configuration of a fiber optic biosensor used to detect a small molecule ground water contaminant, trinitrotoluene (TNT), and the bioremediation bacteria, *Burkholderia cepacia* G4 5223-PR1 (G4).

The "ideal" sensor system has the specificity to distinguish between molecules which are structurally similar, the sensitivity to detect molecules at vanishing small concentrations, the rapidity to give real-time results, the adaptability to detect different analytes, and the simplicity to facilitate real world application.[1] Complete incorporation of all these properties within one device is a daunting, if not an impossible task.

What is possible, however, is a technique which weights each of these properties in accordance to its importance in a given application. The use of immunoassays in conjunction with an evanescent wave fiber optic biosensor is a unique way of balancing the demands for specificity, sensitivity, rapidity, adaptability, and simplicity.

Total internal reflection and the evanescent wave. Light propagating down an optical fiber does so by total internal reflection. [2] If the angle at which the light strikes the internal surface of the fiber is less than the critical angle (Θ_c), then all of the light is reflected as shown in figure 1. If the angle is greater than the critical angle, then some of the light is reflected and some of the light escapes the core of the fiber. The critical angle is determined by the refractive index of fiber core (n_1) and the refractive index of the media surrounding the core (n_2) as shown in equation 1. Using this scheme, optical fibers are able to transmit light over long distances with very little loss of intensity.

$$\Theta_c = \sin^{-1}(n_2/n_1) \qquad (1)$$

Simultaneous with the light's reflection off the internal surface of the fiber is the creation of an electromagnetic field at the external surface of the fiber core. This field extends into the surrounding media and is called the evanescent wave. The intensity of the field decays exponentially as the distance from the surface of the probe increases. The effective distance this field penetrates the external media is less than one wavelength of light [3] and is very sensitive to the incident angle Θ_c, the refractive indexes of the internal media (n_1), and external media (n_2). For the typical optical fiber used in an immunoassay with a fused silica core surrounded by an aqueous media, the penetration depth of the evanescent wave is on the order of 100 nm.

Biosensor Probes. For the fiber optic biosensor used here, a portion of protective cladding on the exterior of the optical fiber is removed from the distal 10 cm of the fiber to expose a core of fused silica. This exposed region becomes the probe. Antibodies are covalently attached to the exposed core. When the probe is in contact with a sample containing an analyte, the immobilized antibody specifically binds the analyte from the bulk solution and concentrates it on the surface of the fiber within the evanescent zone. Any fluorophore associated with the analyte is also immobilized within the evanescent wave. Excitation of the fluorophore by light in the evanescent wave leads to fluorescent emission which generates a detectable signal. Two different methods of associating a fluorophore with the analyte are described below.

The specificity of the biosensor originates from the specificity of the interactions between the analyte and the immobilized antibody. Since both the binding of the analyte and the excitation of the fluorophore are extremely fast, this system can have a very rapid response time. Adapting the biosensor to a different analyte requires changing the immobilized antibody to one which is specific for the new analyte.

Sensitivity of the biosensor is a complex issue. Ultimately, the level of detection is determined by the binding affinity of the recognition element, the ratio of signal-to-noise from both the sample and the hardware, and the sensitivity of the instrument. The two predominate sources of "noise" in the system are: (1) non-specific retention of the fluorophore within the evanescent wave, and (2) stray excitation light.

Fiber Optic Fluorimeter. The function of the fiber optic fluorimeter is to create an evanescent wave on the surface of the probe at a wavelength which will excite the chosen fluorophore, and to detect the fluorescent emission from dye molecules present within the evanescent wave. Figure 2 shows the optical path of the fluorimeter. Notice that near the entrance to the optical fiber, the excitation light originating at the laser and emission light returning from the probe travel the same path. A dichroic mirror and bandpass filter separates the intense laser light, direct and scattered, from the fluorescent emission. The dichroic mirror reflects the low wavelength laser excitation light and transmits the high wavelength emission light. The bandpass filter further defines the light which will be detected as "signal".

Assay Configurations for Environmental Monitoring

The fiber optic biosensor is intentionally designed to operate in a region of the spectra where environmental samples have minimal intrinsic fluorescence. The fluorophores have been selected to excite in this region and have their fluorescence measured with little to no interference from the fluorescence intrinsic to the sample. The method by which the fluorophore is associated with the analyte at the surface of the fiber is primarily a function of the analyte to be detected. In general, large analytes are detected using a sandwich assay wherein the fluorophore is bound to a second antibody, and low molecular weight analytes are detected using a competitive immunoassay in which the fluorophore is bound to an analog of the analyte.

Sandwich Assay. In the sandwich immunoassay, a complex forms which includes the immobilized (unlabeled) antibody, the analyte, and the fluorophore-labeled antibody. The fluorescent signal is directly proportional to the amount of analyte present. For sandwich assays to be effective, the analyte must be large enough to have two distinct, and non-overlapping binding regions: one for binding the analyte to the fiber, and the other for binding the secondary binding molecule. Detection of whole cells is an ideal application of the sandwich assay since whole cells have multiple identical, non-interfering epitopes. This format was used for the detection of G4.

Competitive Binding Assay. In the competitive assay, the analyte and a fluorophore-labeled analog compete for a limited number of available binding sites. The competitive assay is often employed with small molecules lacking two independent epitopes. Thus, the competitive binding assay is ideal for the detection of a small molecule like TNT.

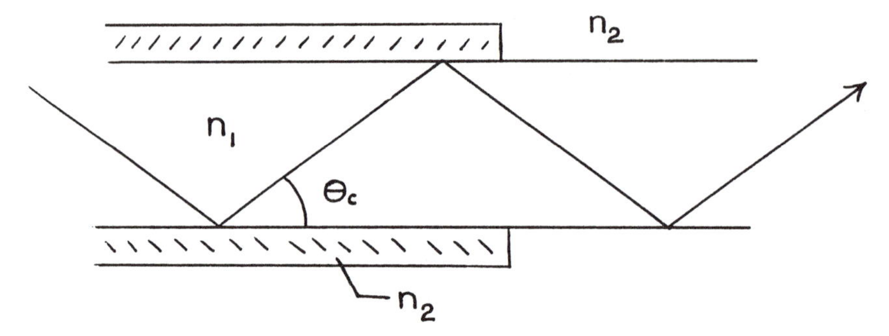

Figure 1. Total Internal Reflection. Cross section of the end of an optical fiber. Refractive index of the core is n_1, and refractive index of either the cladding (hatched) or aqueous media is n_2. Light reflecting at angles less than Θ_c under go total internal reflection.

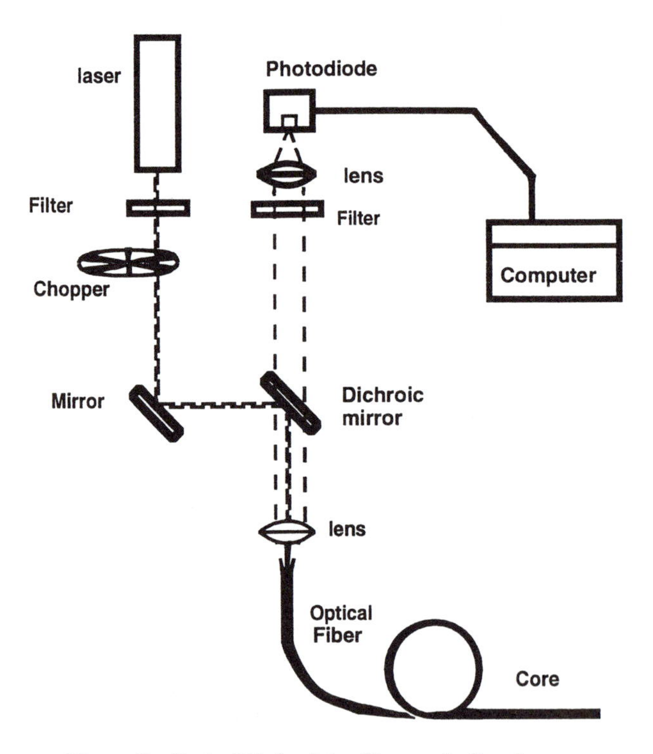

Figure 2. Optical Path of the fiber optic fluorimeter.

With a fiber optic competitive immunoassay, antibodies are immobilized onto the fiber surface and exposed to a solution containing the test sample and a known quantity of the fluorescently-labelled analog. The fluorescent signal is inversely proportional to the analyte concentration. By developing a method to remove the bound analyte or labeled analog ("regenerate" the fiber), we have been able both to accurately calibrate individual fibers for maximum signal and to use one fiber for multiple analysis. Use of a single fiber for multiple analysis decreases variability and expense. A requirement for the success of this assay method, however, is identification of an antibody which has similar binding affinities for the fluorescently-labelled analog and the analyte.

Detection of Whole Bacteria. Microbial bioremediation strategies can address several classes of toxic contamination. One toxic compound that has been the focus of much research is trichloroethylene (TCE).[4] TCE is an industrial degreasing solvent and one of the major ground water contaminants in the United States.[5,6] One organism capable of degrading TCE is the gram negative bacillus, *Burkholderia cepacia* G4 5223-PR1 (G4)[7-9]. The necessity of detecting bacteria used in bioremediation can not be overstated. Applications include: optimization of growth conditions in the laboratory, runs within an off site mesocosm, determination of biological fate, and proof of efficacy. In this study, sandwich immunoassays have been optimized to detect G4 using the evanescent wave biosensor.

For the data shown in figures 3 and 4, the biological binding molecule immobilized on the surface of the fiber is an IgG monoclonal antibody against the O-antigenic region of the lipopolysaccaride that coats the exterior of the G4 cells. This polysaccharide epitope has the advantage that its expression is constant over the lifecyle of the cells, unlike surface proteins which respond readily to the metabolic and reproductive needs of the organism.[7] Characterization of the specificity of this monoclonal by others[7], using ELISA, showed no cross reactivity with 11 strains of bacteria common to water and soil.

Actual construction of a biosensor probe tailored for the detection of G4 is a multistep process. First, the exposed fused silica core of the probe is cleaned with concentrated hydrofluoric acid (HF), then the 200μm core is tapered to 65μm by concentrated HF. Tapering requires about 40 minutes.[10] 3-Mercapto-proply-trimethoxy-silane reacts covalently with the surface hydroxyls to produce immobilized thiol groups. Next, the malimide reactive group of the heterobifunctional crosslinker N-succinimidyl-4-maleimidobutyrate is reacted with the immobilized thiol to yield a tethered succinimide capable of reacting with the primary amines of a protein.[11,12] In the final step, the tethered succinimides react with the IgG (0.05 mg/ml) to covalently bind it to the surface. Antibodies immobilized in this fashion are randomly oriented, but retain 80-100% of their original immobilized binding activity after 12 months in PBS at room temperature .

The antibody-dye conjugate used in the sandwich assay was formed by reacting commercially available Cy3 dye (ex:552, em:565, Biological Detection Systems, Pittsburgh, PA) with IgG in 0.05M NaBorate, 0.05 M NaCl, pH 9.4, for 30 minutes.[13,14] The reaction was stopped and unreacted dye was removed by placing

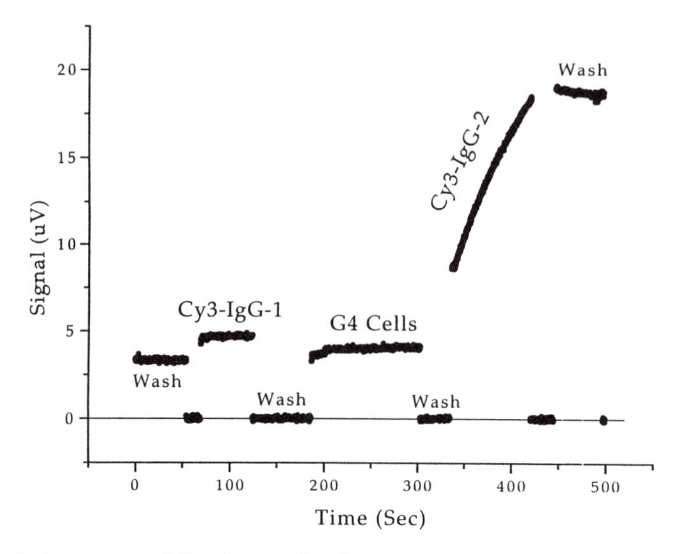

Figure 3. Response of fluorimeter for the course of a single assay. Cy3-IgG-1 and Cy3-IgG-2 are the first and second application of the antibody-dye conjugate, Cy3-IgG, respectfully.

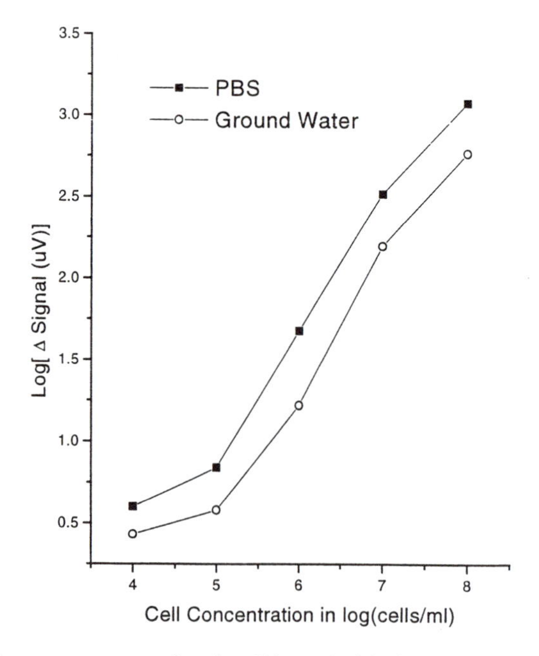

Figure 4. Response curves for the G4 sandwich immunoassay in buffer and ground water, for single fibers. All reagents used for the ground water assay were two months older than those use for the PBS assay. Cells in PBS were exposed to the fiber for 5 minutes and those in ground water were exposed to the fiber for 2 minutes.

the mixture directly on a Biogel P10 1 x 10cm column using pH 7.4 phosphate buffered saline (PBS) as wash buffer. This protocol resulted in crosslinking an average of 4 dye molecules per molecule of IgG as determined by UV/VIS absorbance.

Figure 3 shows the response of fluorimeter over the course of a single assay. Initial "Wash" with pH 7.4 PBS containing bovine serum albumin (2 mg/ml), casein (2 mg/ml), and Triton X-100 (0.1%) measures the ambient rate of photobleaching for the fiber, wash buffer, and instrumental drift. In this figure, there is negligible change in the signal. Closing the shutter causes the signal to drop to zero. Initial exposure to antibody-dye conjugate, "Cy3-IgG-1", measures the fluorescent level due to nonspecific binding of conjugate to the fiber. Although the Cy3-IgG-1 signal is above that of the wash buffer, it does not increase with time. The portion of the curve labeled "G4 Cells" measures signal response during the period the G4 cells were incubated with the fiber and demonstrates that there is very little signal associated with the cells alone. The next "wash" removes cells weakly bound to the fiber. The second application of antibody-dye conjugate, "Cy3-IgG-2", tracks the binding of the Cy3-IgG to cells retained on the fiber. The final "Wash" removes any Cy3-IgG which is associated with, but not bound to the fiber. Ultimately the signal is taken to be the difference between the final wash and Cy3-IgG-1, in uV. Cell concentration was 10^6 cells/ml as determined by acradine orange.

Figure 4 is a standard curve for detection of G4 in PBS and ground water. Each point on the graph was determined by an assay protocol similar to that detailed in figure 3. Here, a sample media of either PBS or ground water was spiked with a known number of G4 cells. This sample was placed on a fiber and allowed to interact with the immobilized IgG for times varying from 1-10 minutes. Unbound cells were removed with a PBS wash buffer containing bovine serum albumin (2 mg/ml), casein (2 mg/ml), and Triton X-100 (0.1%). Next, fluorescently-labeled monoclonal antibodies Cy3-IgG (5ug/ml) bind to the surface of the retained cells for 2-5 minutes. It is the binding of the Cy3-IgG to the retained G4 cells that generates the signal. The fiber is washed to remove any Cy3-IgG non-specifically absorbed to the fiber or the cells. Using the sandwich assay with α-G4-IgG on the fiber and α-G4-Cy3-IgG (5ug/ml) as the secondary antibody approximately 10^4 cells/ml could be detected.

The G4 sandwich assay is extremely robust with regards to reagents and cell sample times. Even though all the reagents used to determine the G4 standard curve in ground water were two months older than those used to determine the G4 standard curve in PBS, and all the samples in PBS were allowed to bind for 5 minutes while those in ground water were allowed to bind for 2 minutes, the basic shapes of the standard curves are almost identical.

Work with real world samples is more complex than work in a PBS media. Each specific application requires a normal compliment of positive and negative experimental controls. Because the specificity of the detection originates from the specificity of the antibody, there is always a potential of false positives due to the existence of a natural epitope similar to that found on the surface of G4. With regard to false negatives, anything which competitively or non-competitively inhibits G4 binding has the potential to skew results. Sources of non-competitive inhibition could be anything causing denaturation of the antibody or a decrease of its binding activity, such as extremes in ionic strength, pH, organic solvent, or protease. Our own work with ground water and surface water from the Potomac river has not been complicated by high signals from "fluorescing" water, false positives, or false negatives.

The search for a more sensitive and accurate detection system motivated the investigation of analyzing rate of signal increase rather than the absolute change in signal. Rate analysis has several advantages over a signal based on change in magnitude. First, the rate signal is less sensitive to errors in Cy3-IgG-2 exposure time. Second, it is very easy to correct for photobleaching and instrument drift. Third, the rate is independent of fluorescent signals which are constant over time. For instance, in magnitude analysis, the jump caused by Cy3-IgG-1 will ultimately decrease the instrument response when it is subtracted from the final wash. In rate analysis, this initial jump need not be considered. Pilot experiments in PBS using rate analysis indicate that this approach results in increase of sensitivity of approximately one order of magnitude (work in progress). At the present, the lower limit of detection is determined more by the fiber-to-fiber variation than nonspecific antibody binding or interfering light.

Detection of Ground Water Contaminants. Many of the ground water contaminants of concern are low molecular weight molecules. These include pesticides, PCB's and explosives such as trinitrotoluene (TNT). As a proof of principle demonstration, a competitive immunoassay was developed for the detection of TNT.

In preparation for the TNT assay, the analog trinitrobenzene (TNB) was fluorescently-labelled with the cyanine dye Cy5 (ex:650, em:670) to form Cy5-TNB. A monoclonal antibody, 11B3 [15], was immobilized on the surface of the probe as the recognition molecule. All buffers, except for the regeneration buffer, contained phosphate buffered saline pH 7.4 with 10% ethanol and 2 mg/ml bovine serum albumin. The regeneration buffer contained 50:50 PBS:ethanol

In the actual assay, the antibody-coated fiber probe was first exposed to 7.5 ng/ml Cy5-TNB in buffer. Then regeneration buffer was used for 1 minute to remove the bound fluorescent analog from the probe. After the regeneration, the antibody-coated probe was equilibrated in buffer and exposed to the test sample containing TNT and 7.5 ng/ml Cy5-TNB.

The cycle of "measurement -- regeneration -- equilibration" was repeated up to 15 times on each fiber with less than 30% loss of antibody activity. Figure 5 shows the detection of several concentrations of TNT with a single fiber. In between samples, the binding activity of the immobilized antibody was calibrated by applying the Cy5-TNB without any TNT present. As shown in figure 6, the limits of detection for TNT with the above protocol are 10 ng/ml. Response of the competitive binding assay is linear over at least two orders of magnitude. Assays performed with spiked 'real world' water samples such as river, harbor, and bilge water showed the same limits and range of detection. In these studies, the water samples contained diverse particulate contaminants and were used unfiltered with dilution for the addition of Cy5-TNB in buffer (1:10 vol:vol).[16]

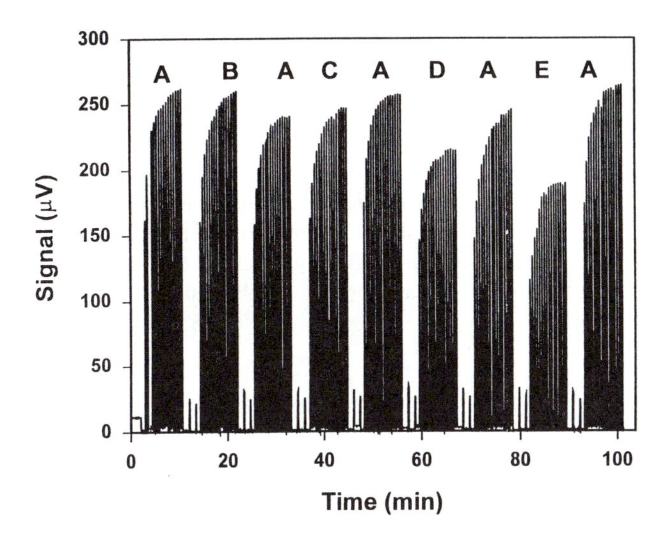

Figure 5. TNT competitive immunoassay on fiber optic biosensor. A single fiber optic probe was exposed to various solutions containing Cy5-TNB(7.5 ng/ml) + TNT (0-50 ng/ml). Concentrations of TNT (ng/ml) were: A=0, B=1, C=5, D=10, E=50.

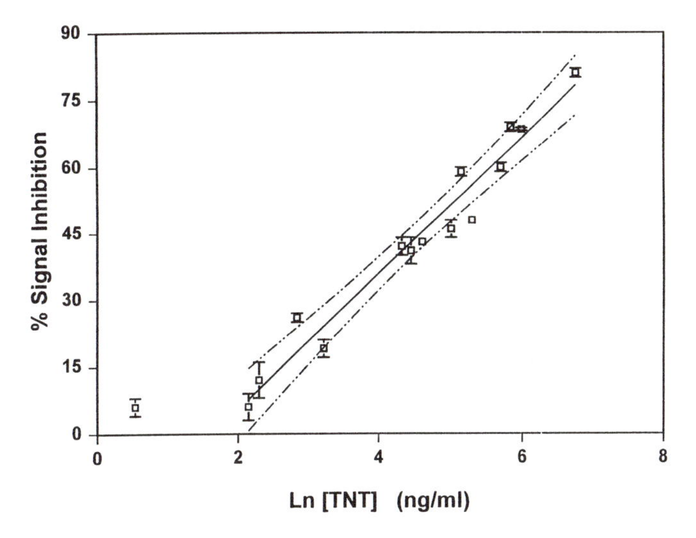

Figure 6. Standard response curve for the TNT competitive immunoassay in buffer. TNT (1-860 ng/ml) solutions were assayed using the fiber optic biosensor. The percent inhibition of the reference signal for Cy5-TNB only for each concentration is shown. The 95% confidence intervals are shown. A minimum of 3 assays were performed for each concentration with the exception of 200 ng/ml TNT.

CONCLUSIONS

The use of immunoassays on the evanescent wave fiber optic biosensor is one way of solving the problem of having a single instrumental platform that can be readily adapted to the detection of a wide variety of analytes. Two widely different analytes discussed here are gram negative bacteria (G4), and a small molecule, trinitrotoluene (TNT). Although each analyte requires a unique set of reagents, the basic steps of forming the fiber probes and the instrumentation are the same. G4 cells were detected at 10^4 cells/ml and TNT at 10 ng/ml (8ppb). In both cases the assays take only minutes to perform.

Acknowledgments

Notice: The U.S. Environmental Protection Agency (EPA), through its office of Research and Development (ORD) partially funded and collaborated in the extramural research described here. It has been subject to the Agency's peer review and approved as an EPA publication. The U.S. Government has the right to retain a non-exclusive, royalty-free licenses in and to any copyright covering this article.

This work was funded in part by the Office of Naval Research / Naval Research Laboratory. The views expressed here are those of the authors and do not represent those of the U.S. Navy, Department of Defense, or the U.S. Government.

We would like to thank the U.S. Environmental Protection Agency Environmental Research Laboratory, Gulf Breeze (GBERL) for the G4 strain and Jorg Winkler for his generous gift of α-G4 monoclonal antibodies.

References

1. Alvarez-Icaza, M.; Bilitewski, U. *Anal. Chem.* **1993**, *65*, 525.
2. Axelrod, D.; Burghardt, T. P.; Thompson, N. L. *Ann. Rev. Biophys. Bioeng.* **1984**, *13*, 358.
3. Anderson, G. P.; Golden, J. P.; Cao, L. K.; Wijesuria, D.; Shriver-Lake, L. C.; Ligler, F. S. *IEEE Eng.Bio. Med.* **1994**, *13*, 358.
4. Folsom, B. R.; Chapman, P. J.; Pritchard, P. H. *Appl. Environ. Micro.* **1990**, *56*, 1279.
5. Rajagopal, R. *Environ. Prof.* **1986**, *8*, 244.
6. Love, O. T.; Eilers, R. G. *J. Am. Water Works Assoc.* **1982**, *80*, 413.
7. Winkler, J.; Timmis, K. N.; Snyder, R. A. *Appl. Environ. Micro.* **1995**, *61*, 448.
8. Sheilds, M. S.; Reagin, M. J. *Appl. Environ. Micro.* **1992**, *58*, 3977.
9. Sheilds, M. S.; Montgomery, S. O.; Cuskey, S. M.; Chapman, P. J.; Pritchard, P. H. *Appl. Environ. Micro.* **1991**, *57*, 1935.
10. Golden, J. P.; Shriver-Lake, L. C.; Anderson, G. P.; Thompson, R. B.; Ligler, F. S. *Opt. Eng.***1992**,*31*, 1458.

11. Bhatia, S. K.; Shriver-Lake, L. C.; Prior, K. J.; Georger, J. H.; Clavert, J. M.; Bredehorst, R.; Ligler, F.Anal.*Biochem.* **1989**, *178*, 408.
12. Ligler F.S. (misspelled Eigler); Georger, J. H.; Bhatia, S. K.; Calvert, J.; Shriver-Lake, L. C.; Bredehorst, R. *U.S. Patent # 5,077,210* **1991.**
13. Southwick, P. L.; Ernst, L. A.; Tauriello, E. W.; Parker, S. R.; Mujumdar, R. B.; Mujumdar, S. R.;Clever, H. A.; Waggoner, A. S. *Cytometry* **1990**, *11*, 418.
14. Smith, L. M. *Nucleic Acid Research* **1985**, *13*, 2399.
15. Whelan, J. P.; Kusterbeck, A. W.; Wemhoff, G. A.; Bredehorst, R.; Ligler, F. S. *Anal. Chem.* **1993**,*65*, 3561.
16. Shriver-Lake, L. C.; Breslin, K. A.; Charles, P. T.; Conrad, D. W.; Golden, J. P.; Ligler`, F. S.*Anal. Chem.* **1995**, *in press*,

RECEIVED September 15, 1995

Chapter 5

A New Method for the Detection and Measurement of Polyaromatic Carcinogens and Related Compounds by DNA Intercalation

John J. Horvath, Manana Gueguetchkeri, Adarsh Gupta, Devi Penumatchu, and Howard H. Weetall

Biotechnology Division, National Institute of Standards and Technology, Gaithersburg, MD 20899–0001

A simple and sensitive method for detection and quantification of carcinogens and related polyaromatic compounds by DNA intercalation has been devised. The experimental technique is based on the phenomenon of DNA intercalation using fluorescence polarization for quantitative measurements. The assay architecture is analogous to a protocol presently used for competitive immunoassays, whereby an intercalating dye or fluorochrome would compete with or be displaced by the analyte (test compound). Our competitive binding assay utilizes DNA binding sites and a fluorescent intercalating dye. The carcinogenic analytes investigated were benzo[a]pyrene, dibenz[a,h]anthracene, dibenz[a,j]anthracene, benz[a]anthracene, benzo[j]fluoranthene, benzidine, aniline, parathion, pentachlorophenol and nitrobenzene. The noncarcinogenic hydrocarbons studied were naphthalene, anthracene, phenanthrene, benzo[k]fluoranthene and 1,2,3,4,5,6,7,8-octahydronaphthalene. The intercalating dyes examined were acridine orange, ethidium bromide, proflavin and 4,6-diamidino 2-phenylindale chloride (DAPI). Of the compounds studied, only the single ring compounds gave a negative response. All the other compounds investigated displaced the intercalated dye molecule. It has been determined that molecules require at least two adjacent benzene rings for intercalation to occur. The detection limits of this analytical method were between 10^{-5} mol/L and 10^{-8} mol/L for all materials tested.

Intercalation is a reversible insertion of a guest species into a lamellar host structure. Study of the reactions between guest molecules and the host molecule (double stranded DNA) has been ongoing since 1947, when Michaelis[1] observed and correlated dramatic changes in the visible absorption spectra of basic dyes when binding to DNA. Quantitative binding studies were made by utilizing equilibrium dialysis[2,3], thermodynamic models such as Scatchard plots[4], viscosity[5], NMR[6] and fluorescence spectroscopy[7-10].

The intercalative interaction of dyes with DNA were intensively studied and characterized by using many different methods[11-18]. In addition to dyes, other compounds such as aminoquinolines[19] fused aromatics such as diamino-phenyl indoles,[20] a large number of polycyclic aromatic hydrocarbons[21] and benzopyrenediol epoxide[22] also intercalate into the DNA. It is well known that acridine orange becomes brightly fluorescent when bound to double stranded DNA.[23] In the present work, this property of acridine orange has been used for the rapid detection and quantitation of carcinogens and related compounds which intercalate into the DNA. The assay architecture is analogous to a protocol generally used for competitive immunoassay, whereby an intercalating dye or fluorochrome competes for a binding site or is displaced by the analyte. The experimental technique is based on the phenomenon of DNA intercalation using fluorescence polarization for quantitative measurements. The initial guide for this study was a U.S. patent by Richardson and Schulman,[24] who used the classic intercalators acridine orange, ethidium bromide, and proflavin with calf thymus DNA to measure small quantities of the drug actinomycin D. Our competitive binding assay utilized DNA/acridine orange as a competitive agent to the intercalating test compound. The uniqueness of this method is that any test molecule, fluorescent or not, which binds to DNA, is detected by the displacement of the fluorescent intercalator. The action is monitored by using excitation and emission wavelengths specific for the intercalator. In most cases, the fluorescence of test compounds will not interfere if they do not strongly overlap with the chosen detector dye.

The initial goal of this research was to investigate the feasibility of using fluorescence and DNA intercalation to develop a rapid, simple and sensitive method for the detection and quantitation of several varieties of airborne carcinogens. Present methods recommended by the EPA for the collection and analysis of airborne carcinogens require 24 hour air sampling times plus analysis time. Analyses typically utilize gas chromatography/mass spectroscopy or gas chromatography/electron capture detection for identification and quantification.

Materials and Methods

Chemicals and their Preparation. The first chemical carcinogens to be examined by the DNA/Intercalation fluorescence techniques were typical carcinogens of several classes which were of interest for possible monitoring applications. Aniline, benzidine, benzo[a]pyrene, parathion, nitrobenzene, benz[a]anthracene, benzo[j]fluoranthene, benzo[k]fluoranthene and dibenz[a,j]anthracene were procured from the NCI chemical carcinogen repository* (Kansas City, Missouri). Pentachlorophenol, dibenz[a,h]anthracene, the fluorescent intercalating dyes and other hydrocarbons used in these studies were obtained from Sigma Chemical Company (St. Louis, Missouri). The 1,2,3,4,5,6,7,8-octahydronaphthalene was obtained from Aldrich Chemical Company (Milwaukee, Wisconsin).

We prepared the stock solutions of acridine orange, proflavin, ethidium bromide, aniline, benzidine and DAPI in a standard buffer containing 8 mmol/L Tris, 50 mmol/L NaCl and 1 mmol/L EDTA, pH 7.0. The stock solutions of naphthalene, anthracene, phenanthrene, parathion, pentachlorophenol, nitrobenzene, naphthalene, benzo[a]pyrene, dibenz[a,h]anthracene, dibenz[a,j]anthracene, benz[a]anthracene, benzo[j]fluoranthene, benzo[k]fluoranthene, and 1,2,3,4,5,6,7,8-octahydronaphthalene were prepared in ethanol. The concentration of the stock solutions for the carcinogens and hydrocarbons was between 1×10^{-3} mol/L and 5×10^{-5} mol/L. We carried out serial dilutions with the standard buffer before the experimental measurements. Serial dilutions of the stock solutions were prepared in ethanol using standard buffer containing 5% ethanol.

Nucleic Acids. We dissolved calf thymus DNA in the standard buffer by placing the DNA and buffer in a screw-capped vial and agitating for at least 12 hours. The concentration of DNA in solution was determined by absorption at 260 nm in a 1 cm path length quartz cuvette (1.00 Absorption units of duplex DNA was assumed equal to 50 μg/mL of DNA in solution).[25]

Instrument and Polarization Measurement. We used an SLM 8000C scanning spectrofluorometer, manufactured by SLM Aminco Instruments, Inc., Urbana, Illinois, for all fluorescence measurements. This instrument uses a 450 W Xenon arc lamp as the excitation source and a double grating monochromator for the selection of the excitation wavelength. A single grating monochromator monitored the fluorescence. The sample was contained in a 1 cm path length quartz cuvette with excitation and emission slits normally set at 8 nm and 4 nm bandpass, respectively. The normal 90° fluorescence geometry was used. The excitation and emission paths contained adjustable Glan-Taylor polarizers. Photomultipliers monitored both reference and signal channels and used analog signal processing. A depolarizer was used on the lamp output beam (the lamp showed partial polarization). The excitation lamp was monitored by a quantum counter (Rhodamine B), and fluorescence signals were measured and normalized by this reference signal in order to minimize the effects of lamp drift.[2]

---

* Certain commercial equipment, instruments and materials are identified in this paper in order to specify adequately the experimental procedure. In no case does such identification imply recommendation or endorsement by NIST, nor does it imply that the material or equipment is necessarily the best available for that purpose.

Both excitation and emission polarizers could be adjusted to transmit either vertically (0°) or horizontally (90°) polarized light. The two monochromators had different transmission efficiencies for the vertically and horizontally polarized light that modified the actual measured intensities. This difference in sensitivity is commonly referred to as the G factor and must be determined in order to obtain polarization measurements. We employed the method of Lakowicz[26] to obtain the G factor and to calculate polarization values.

Determination of G. We utilized a dilute suspension of glycogen as a scattering sample. We applied polarized excitation light, and the scattered light was 100% polarized. When a high concentration of glycogen was used, multiple scattering led to decreased polarization values. Horizontally polarized light (H) was used for excitation, and the scattered light was measured with vertical (I_{HV}) and horizontal (I_{HH}) polarizer positions. The G value (system polarization response) can be described as:

$$\frac{I_{HV}}{I_{HH}} = G \tag{1}$$

We then used the G value to obtain the actual values of the parallel (I_{vv}) and perpendicular (I_{HV}) intensities, unbiased by the detection system.

Sample Measurement. To measure the polarization of a sample after obtaining the G value, vertically polarized excitation light at a fixed wave-length, determined by the dye (490 nm for acridine orange), was used and the fluorescence emission (530 nm) was first measured with the emission polarizer at 0°(V), and then 90°(H). Data were obtained for 30 seconds with 1 second integration intervals at each polarization setting. Obtaining the average of the center 25 (25 pts) seconds of each yields signals I_{VV}, I_{VH} (vertical excitation, V or H emission). These values are used with the G value to obtain the unbiased parallel intensities

$$\left(\frac{I\|}{I\perp}\right)$$

$$\frac{I_{VV}}{I_{HV}} \frac{1}{G} = \frac{I\|}{I\perp} \bullet \tag{2}$$

We then calculate the anisotropy:

$$r = \frac{(I_{\parallel} / I_{\perp}) - 1}{(I_{\parallel} / I_{\perp}) + 2} \quad (3)$$

The anisotropy is used to calculate the polarization

$$p = \frac{3r}{2 + r} \quad (4)$$

Once the G value was obtained, we calculated the equations using the observed intensities as inputs, obtaining polarization values as outputs along with the averages and standard deviations of the signals.

All the measurements were made in a 1 cm path length quartz cuvette in the spectrofluorometer. We added a measured quantity of acridine orange into the buffer contained in the cuvette. The fluorescence polarization of this solution was then measured. To the same cuvette, we added a known quantity of DNA. We mixed the solutions in the cuvette for 5 minutes and measured the polarization. The increase in dye polarization is due to DNA intercalation, which stops the free rotation of the dye molecules. However, other possible mechanisms for interactions with DNA, which include non-specific electrostatic binding and groove-binding interactions, can also increase the observed polarization. We then added test compounds to the cuvette in successively increasing amounts. We continuously stirred the solution using a magnetic stirrer and stir bar placed in the cuvette. We continued the polarization measurements until the cuvette was filled with liquid (3.8 ml). The initial acridine orange concentrations in the cuvette were approximately 4.0×10^{-7} mol/L, and when filled, 2.2×10^{-7} mol/L.

Results

Determination of the Mechanism for the Observed Polarization Changes.

Upon addition of DNA to acridine orange, the observed polarization increases due to intercalation. Another possible mechanism for this increase could be due to non-specific electrostatic binding to the DNA. To determine the amount of acridine orange electrostatically bound to the DNA, we performed a series of experiments using ethidium bromide as the test intercalator at different sodium chloride concentrations. The electrostatically bound acridine orange will be a function of salt concentration, while the intercalated acridine orange will not be affected. The polarization of the free acridine orange did not change with increasing salt concentration; however, the polarization of the acridine orange-DNA complex decreased, as shown in figure 1.

This curve is similar to the one observed for the binding of proflavine to DNA, as a function of competitor salt concentration, as determined by equilibrium dialysis[27].

As the salt concentration increased, the electrostatically bound acridine orange was displaced from the DNA, decreasing the observed polarization value. As shown in figure 2, at low concentrations of ethidium bromide, the initial polarization value decreased with increasing salt concentration, and remained relatively constant at low ethidium bromide concentrations. Once the detection limit is reached, the polarization decreases linearly with ethidium bromide concentration. In figure 3, a normalized version of figure 2, the detection limit (LOD), the point where the polarization starts to decrease, is shown to be unaffected by salt concentration.

In the flat polarization region (at low ethidium bromide concentrations), the ethidium bromide molecules are most likely filling intercalation sites not containing acridine orange. Therefore, no change in polarization is observed. At the concentration exceeding the detection limit, the ethidium bromide starts to replace intercalated acridine orange. From this point on, the change in polarization is totally due to the ethidium bromide, a known intercalator[28], displacing intercalated acridine orange.

Measurements of Carcinogens by Displacement of Acridine Orange. The carcinogens examined were benzo[a]pyrene, dibenz[a,h]anthracene, dibenz[a,j]-anthracene, benz[a]anthracene, benzo[j]fluoranthene, benzidine, aniline, nitrobenzene, parathion, and pentachlorophenol. Figure 4 presents the results of the fluorescence polarization measurements for benzo[a]pyrene using acridine orange as the intercalating fluorescent indicator. The data points indicate the polarization of the sample after the addition of the test compound (carcinogen) at different concentrations. Initially, the polarization value remained constant at low concentrations. The detection limit in this system is indicated by an arrow on the plot; after this concentration, the observed polarization begins decreasing with increasing concentration of the test sample. As shown in Figure 4, the LOD for benzo[a]pyrene is 4.7×10^{-8} mol/L with a relative standard deviation of 4%.

Benzidine, a compound with a biphenyl-ring, indicated no dye displacement (no polarization change) at all concentrations. Aniline, nitrobenzene, pentachlorophenol, and parathion showed similar behavior as that of benzidine, demonstrating that they also could not displace acridine orange from the double stranded DNA. These molecules that did not displace the dye contained a single benzene ring, whereas the acridine orange contained three fused benzene rings with small substituent groups.

To further determine whether single ring molecules could intercalate into DNA on its own, we examined aniline, a fluorescent carcinogen. Using an excitation wavelength of 285 nm and emission wavelength of 334 nm, we monitored the fluorescence of aniline as DNA was added to the cuvette. No change was observed in the aniline fluorescence polarization, even at high DNA concentrations. This indicates that aniline, and most likely benzidine, parathion, pentachlorophenol and nitrobenzene, are not intercalators, or are unable to displace acridine orange from the DNA. Normalized analytical plots for the five carcinogens demonstrating intercalation are shown in Figure 5.

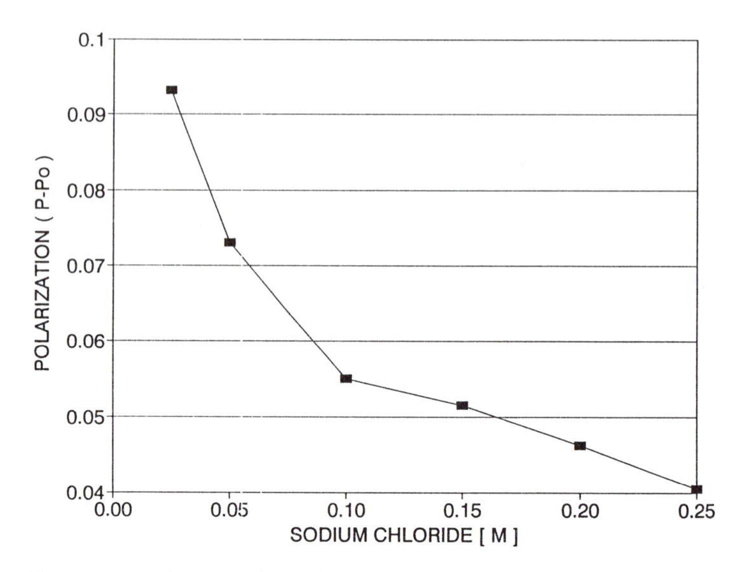

Figure 1. Reduction of acridine orange polarization as a function of sodium chloride concentration. P_o = polarization of free acridine orange (1.0×10^{-6} M), P = polarization after addition of 1.6 μg of calf thymus DNA.

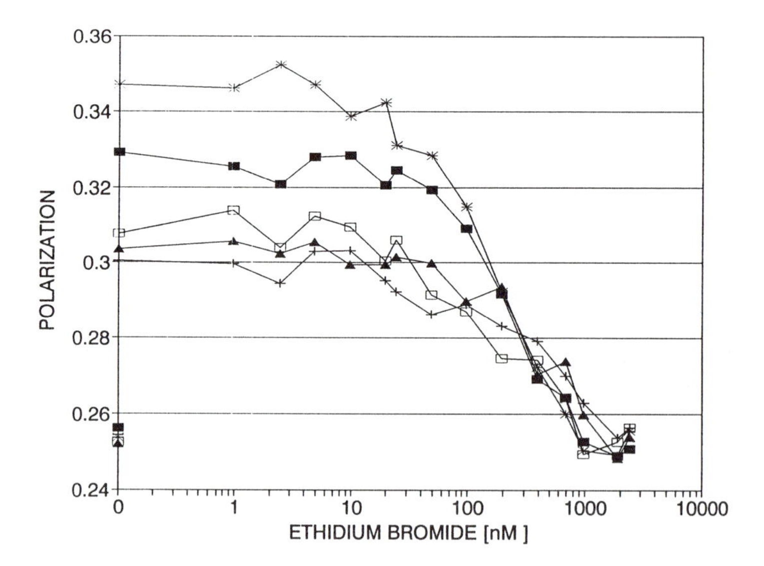

*Figure 2. Analytical plots for Ethidium Bromide at various NaCl concentrations. Acridine orange = 1.0×10^{-6} M, 1.6 μg calf thymus DNA. Sodium Chloride concentrations, * = 0.025 M, ■ = 0.05 M, □ = 0.10 M, ▲ = 0.15 M, + = 0.20 M.*

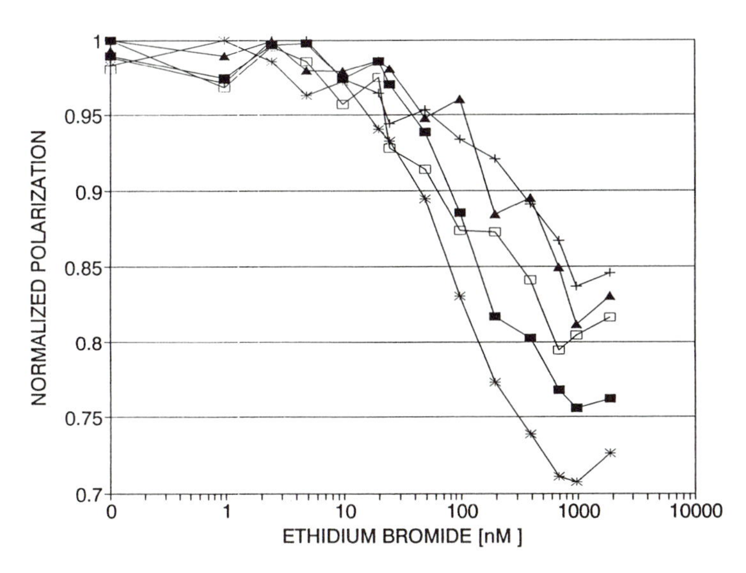

Figure 3. Normalized analytical plots for Ethidium Bromide at various sodium chloride concentrations indicating no change in the limit of detection. Acridine orange = 1.0 x 10^{-6} M, 1.6 μg calf thymus DNA. Sodium Chloride concentrations, ✶ = 0.025 M, ■ = 0.05 M, □ = 0.10 M, ▲ = 0.15 M, + = 0.20 M.

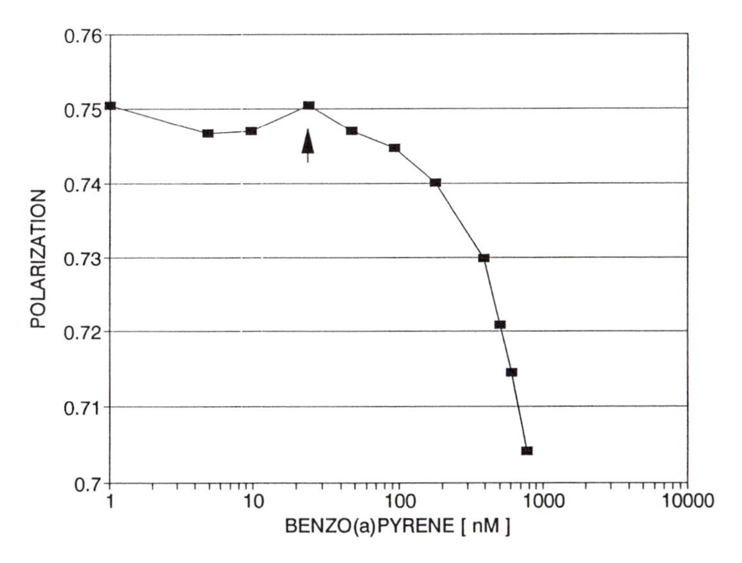

Figure 4. Shows polarization of acridine orange vs log of Benzo[a]pyrene concentrations with calf thymus DNA. Arrow indicates the limit of detection for benzo[a]pyrene. Acridine orange = 5.0 x 10^{-7} mol/L, DNA = 6.4 μg.

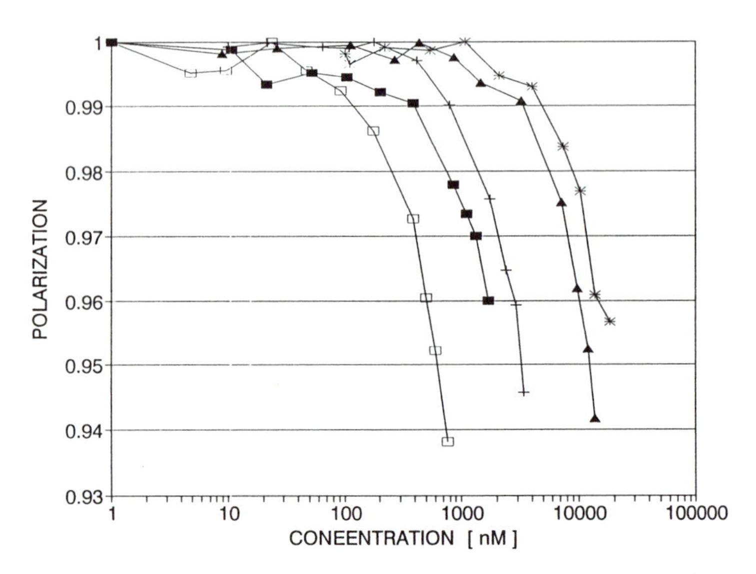

Figure 5. Analytical plots for Carcinogens in Classes III,IV and V. Acridine orange = 5.0 x 10^{-7} mol/L, DNA = 6.4 μg. ✶ = Benzo[a]anthracene (Cl III), ■ = Benzo[j]fluoranthene (Cl IV), □ = Benzo[a]pyrene (Cl V), ▲ = Dibenz[a,j]anthracene (Cl III), + = Dibenz[a,h]anthracene (Cl V).

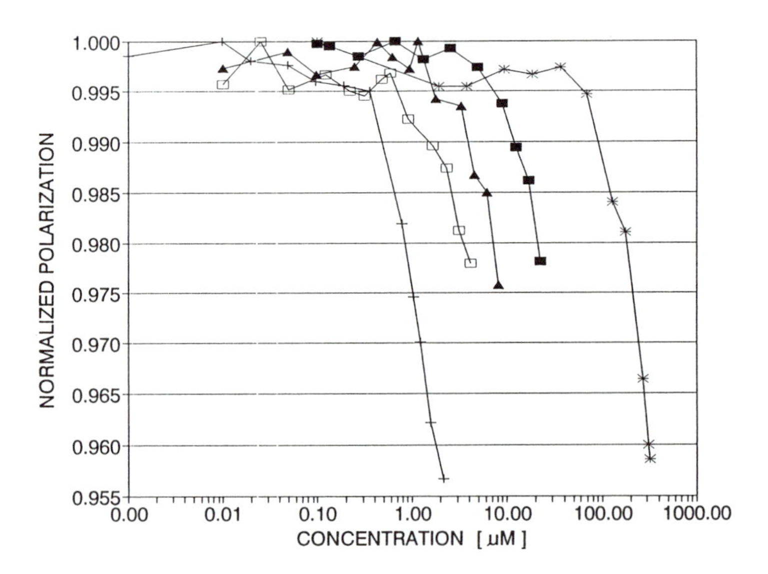

Figure 6. Analytical plots for non-carcinogenic Class I molecules studied. Acridine orange = 5.0 x 10^{-7} mol/L, DNA = 6.4 μg. ✶ = Naphthalene, ■ = Anthracene, □ = Phenanthrene , ▲ = 1,2,3,4,5,6,7,8-octahydronaphthalene, + = Benzo[k]fluoranthene.

Investigations on Criterion for Competition Against Acridine Orange. After observing the results with molecules containing a single benzene ring, we focused the studies on finding the molecular configuration required for competition or displacement of the acridine orange intercalated in the DNA. To investigate in more detail the number of adjacent rings required in intercalating molecules, we examined a series of cyclic hydrocarbons of differing size, structure and aromaticity. Naphthalene, anthracene, benzo[k]fluoranthene, and 1,2,3,4,5,6,7,8-octahydronaphthalene were tested in the calf thymus DNA/acridine orange system. Their minimum detectable concentrations are listed in Table I and the analytical plots are shown in Figure 6.

We also examined the intercalating dyes, ethidium bromide, proflavin and DAPI, by comparing them with acridine orange, yielding LOD's of 6.4×10^{-9} mol/L, 5.8×10^{-9} mol/L and 7.0×10^{-8} mol/L, respectively. The first two dyes containing three-ring systems showed one order of magnitude greater detectability over the two-ring based dye DAPI.

Discussion

Our initial studies validated the concept of using DNA intercalation monitored by fluorescence polarization for detection and measurement of polyaromatic carcinogens and other chemicals containing at least two fused benzene rings. LODs for benzo[a]pyrene and dibenz[a,h]anthracene were 4.7×10^{-8} mol/L (6 ppb) and 1.8×10^{-7} mol/L (25 ppb), respectively. A large difference was observed in the LODs of the carcinogens and noncar-cinogens tested. For example, the LOD for the highly carcinogenic benzo[a]pyrene was 4.7×10^{-8} mol/L, and for the noncarcinogenic naphthalene, the LOD was found to be 3.7×10^{-5} mol/L.

Single-ring compounds, including aniline, 2,4-dinitrophenol, nitrobenzene, parathion and pentachlorophenol, did not intercalate, indicating that a minimum structure of two or more adjacent rings is necessary for intercalation. For intercalation and displacement to occur, the binding strength or affinity of the competing agent for the intercalation site in the DNA must reach a concentration such that it can displace the intercalated dye. The greater the binding constant of the competitive agent, the smaller the required concentration for displacement. A molecule forms no covalent bonds with the DNA upon intercalation, but after insertion between successive base pairs, it is held in place by electrostatic or Van der Waal's forces.[28,29] An intercalating molecule must have sufficient interaction energy to force the stacked base pairs apart to accommodate the intercalator into the DNA.

The results for benzo[a]pyrene, with an LOD approximately one order of magnitude lower than dibenz[a,h]anthracene, are in agreement with the findings of Lesko et al.[30] and Craig and Isenberg,[31] who have also reported stronger binding to DNA for benzo[e]pyrene and benzo[a]pyrene compared to dibenz[a,h]anthracene. Both benzo[a]pyrene and dibenz[a,h]anthracene are known intercalators, and are larger molecules than acridine orange (five benzene rings vs. three). The greater sensitivity of benzo[a]pyrene indicates a stronger intercalation interaction with DNA for the displacement of acridine orange than the dibenz[a,h]anthracene. Benzo[a]pyrene is more compact than dibenz[a,h]anthracene, resulting in a lower LOD.

The size criterion is a very important factor in determining the ability of a molecule to intercalate into DNA.[31-34] When comparing various polycyclic aromatic hydrocarbons (PAH) binding to native DNA, Isenberg et al.[32,33] and Craig and Isenberg[31] proposed that appreciable binding to DNA occurs only if the dimensions of the PAH molecules are small enough to allow the faces of the intercalated molecule to be protected by the DNA base pairs from contact with aqueous media. Other important factors involved in intercalation include the total surface area of the PAH molecule[34,35] and the magnitude of Van der Waals' attractive energies between the intercalated molecules and the adjacent base pairs.[36] Both of these factors are predicted to favor larger binding constants for larger PAH molecules.[34] The lower LOD of benzo[a]pyrene compared to dibenz[a,h]anthracene is due to its more compact structure, allowing more of the intercalated molecule to be protected from the destablizing influence of the aqueous media.

The inability of aniline, benzidine, pentachlorophenol, nitrobenzene and parathion to be intercalated into the DNA may be explained by the fact that these molecules are too small to generate an interaction energy large enough to separate the base pairs, or that it is difficult for a molecule containing a single ring to displace a molecule containing three rings such as acridine orange. We observed no change in the fluorescence polarization for these molecules, indicating a failure to replace acridine orange at the experimentally used concentrations. When we monitored the fluorescence of aniline while adding DNA, we observed no change in polarization, which demonstrates that no intercalation of aniline with DNA occurred up to a concentration of 3.9×10^{-4} mol/L. All the fluorescent intercalating dyes presently used for nucleic acid staining contain two or more fused rings.[37] Based upon this evidence and our experimental results, we conclude that molecules containing a single benzene ring will not intercalate into or displace acridine orange from the DNA, because their small size will not create a large enough interaction energy (electrostatic, Van der Waals, etc.) to displace the indicator molecule.

We studied the size criteria for competitive intercalation by comparing the hydrocarbons naphthalene, anthracene, phenanthrene, benzo[k]fluoranthene and 1,2,3,4,5,6,7,8-octahydronaphthalene, which present a progression from two rings to five rings. The LODs for these molecules is shown in Table I. Naphthalene, with two adjacent rings, showed the poorest detection limit. Anthracene, with three adjacent rings, similar to acridine orange, showed a stronger interaction by displacing at lower concentrations. Phenanthrene also showed an improved LOD over anthracene. In comparing naphthalene and anthracene, we observe along the major axis of the molecule a size increase, which leads to an improvement in detection. The better LOD of phenanthrene compared to anthracene could be attributed to the more compact ring structure of phenanthrene. These observations on the effects of size and compactness agree with earlier observations on polycyclic aromatic hydrocarbons.[30-33] These investigators found that size and conformation played a large role in intercalation "binding."

One possible measure of relative size could be given by molecular weight. In Table II, the molecules that demonstrated intercalation in these studies are listed by their carcinogenicity class, LODs and molecular weights. Figure 7 shows a plot of LOD vs. molecular weight for the molecules listed in Table II. This plot indicates that higher

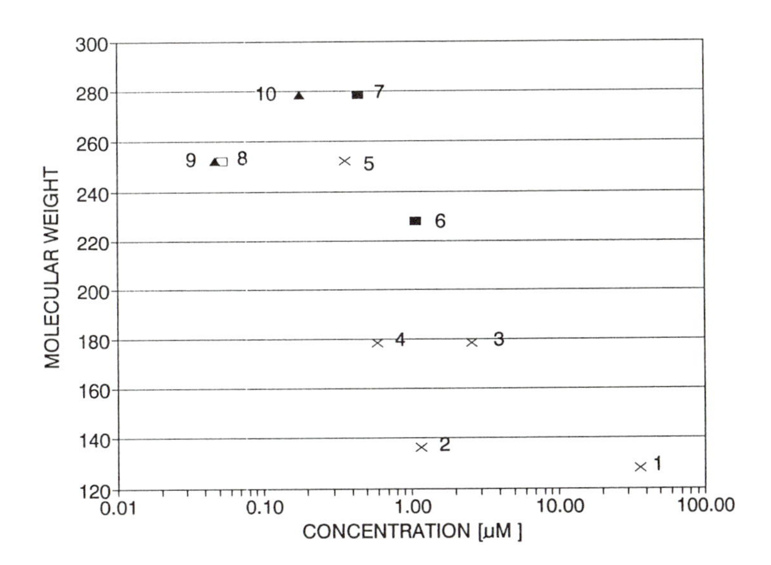

Figure 7. Plot of Molecular Weight vs Limit of Detection for all the intercalating molecules in this study. The numbers correspond to the molecule numbers in Table II. X = Class 1, ■ = Class 2, □ = Class 4, ▲ = Class 5.

Table I : Limits of Detection (LOD) for Non-Carcinogenic Class I molecules

Molecule	LOD(mol/L)
Naphthalene	3.7×10^{-5}
1,2,3,4,5,6,7,8-octahydronaphthalene	1.2×10^{-6}
Anthracene	2.6×10^{-6}
Phenanthrene	5.9×10^{-7}
Benzo[k]fluoranthene	3.6×10^{-7}

Table II : Detection Limits (LOD) and Molecular Weights for All Intercalating Molecules Studied

Name	LOD mol/L	Molecular Weight
Class I		
Naphthalene	3.7×10^{-5}	128.16
1,2,3,4,5,6,7,8-octahydro-naphthalene	1.2×10^{-6}	136.16
Anthracene	2.6×10^{-6}	178.22
Phenanthrene	5.9×10^{-7}	178.22
Benzo[k]fluoranthene	3.6×10^{-7}	252.32
Class III		
Benz[a]anthracene	1.1×10^{-6}	228.28
Dibenz[a,j]anthracene	4.9×10^{-7}	278.32
Class IV		
Benzo[j]fluoranthene	5.3×10^{-8}	252.32
Class V		
Benzo[a]pyrene	4.7×10^{-8}	252.30
Dibenz[a,h]anthracene	1.8×10^{-7}	278.35

molecular weight compounds tend to have lower LODs, although structural and chemical properties also play a major role.

Comparing the lowest molecular weight compound in class I naphthalene (128.16), with the highest one, benzo[k]fluoranthene (252.32), the LOD is reduced by two orders of magnitude, demonstrating the effect of increasing size. The improved LOD of phenanthrene over anthracene, both at the same molecular weight, can be explained by its greater compactness, thereby allowing for a better fit into the DNA. However as compactness increases the hydrophobic driving force will decrease. The observed LOD will be determined by the interactions between the various forces and the physical structure of the molecule.

Two molecules with the same size and structure based on the carbon atoms, naphthalene and 1,2,3,4,5,6,7,8-octahydronaphthalene, exhibit greatly differing LODs, 3.7×10^{-5} mol/L and 1.2×10^{-6} mol/L, respectively. Naphthalene, with a molecular formula of $C_{10}H_8$, is a planar molecule, while the octahydronaphthalene ($C_{10}H_{16}$) is a bent molecule due to its higher degree of saturation, possibly yielding a larger effective volume for interaction with the DNA. The high degree of saturation will also increase the hydrophobicity of the octahydronaphthalene, strengthening its interaction with the DNA base pairs to protect it from contact with the aqueous environment.[30-32]

The LODs obtained for the Class 5 carcinogens benzo[a]pyrene (4.7×10^{-8} mol/L) and dibenz[a,h]anthracene (1.8×10^{-7} mol/L), when compared to the non-carcinogenic naphthalene (3.7×10^{-5} mol/L), indicate a possible relationship between the LOD values obtained in this "assay" method and the carcinogenicity of the molecule? Table II shows the LODs of the molecules studied, as determined by acridine orange displacement, versus carcinogenicity[38] (class I - non-carcinogenic to class V - most carcinogenic). Although it is impossible to make definitive statements on the carcinogenicity/LOD question on this small a sample set, several observations can be made regarding the intercalation mechanism.

Benzo[j]fluoranthene, which only differs from benzo[k]fluoranthene by having one benzene ring in a slightly different position, has a better LOD of approximately one order of magnitude, but is a class IV carcinogen, whereas benzo[k]fluoranthene is a noncarcinogenic Class I molecule. It would also appear that benzo[j]fluoranthene is a more compact molecule, which should enhance intercalation.[30]

For the five-member ring compounds - (benzo[k]fluoranthene, benzo[j]fluoranthene, dibenz[a,j]anthracene, dibenz[a,h]anthracene and benzo[a]pyrene), the most compact, benzo[a]pyrene, also showed the best LOD at 4.7×10^{-8} mol/L. As the structure becomes less compact, the LOD increases. Dibenz[a,h]anthracene, with an LOD of 1.8×10^{-7} mol/L, is approximately three times lower than dibenz[a,j]anthracene by having a benzene ring moved from the h to j position. The small difference in LODs between dibenz[a,h] and [a,j]anthracene, two molecules with different carcinogenicities, again indicate that other chemical interactions are responsible for the carcinogenic ranking, rather than a physical disruption based on DNA intercalation.

These results show that binding affinity as measured by competitive displacement of acridine orange from DNA is not a major criterion for carcinogenicity. The causes of carcinogenicity lie elsewhere. In this study, we find no direct correlation between carcinogenicity and LODs as measured with acridine orange displacement.

To determine the presence and relative concentrations of a competitive intercalating agent, the agent must displace the indicator dye molecule. To obtain the lowest LOD for the competitive agent, one requires an easily displaced dye without any base pair specificity to fill the maximum number of binding sites. Acridine orange, the indicator dye used in these studies, may not be the optimum choice for each competitive agent tested. The other fluorescent dyes that require further examination include ethidium bromide (6.4×10^{-9} mol/L), proflavin (5.8×10^{-9} mol/L) and DAPI (3.5×10^{-8} mol/L). The values indicated are the detection limits for the displacement of acridine orange with these dyes using present experimental protocols. Wilson et al., [20,39] Tanious et al., [15] and Larsen et al., [40] have shown a sequence dependent binding mode for DAPI. It is possible that DAPI may fill additional empty sites in DNA, for which it has strong affinity (AT clusters), before it begins displacing acridine orange. Proflavin, on the other hand, does not have any base pair specificity.[7] Ethidium bromide also shows little or no base pair specificity.[37] DAPI may also have a higher LOD due to its structure, a two-ring indole connected to a single benzene type ring.

We have demonstrated that polycyclic aromatic hydrocarbons can be rapidly measured by using the DNA intercalation -fluorescence technique. The limits of detection obtained are sensitive enough to be used for measurements on environmental carcinogens[41]. Future studies will utilize evanescent wave technology with fiber optics and immobilized DNA to develop a portable real time instrument for in situ measurements of carcinogens..

Conclusions

The DNA intercalation - fluorescence polarization technique was demonstrated to be a simple and rapid method for detection and measurement of ppb quantities of carcinogenic polycyclic aromatic hydrocarbons. Molecules containing a single benzene ring were unable to intercalate into calf thymus DNA because at least two fused rings were required for intercalation to occur. In a series of carcinogenic and noncarcinogenic polycyclic aromatic hydrocarbons no systematic relationship between carcinogenicity and intercalation was observed. The intercalating behavior of polycyclic aromatic hydrocarbons was shown to be a complex function of size, structure and other undetermined variables. The use of different indicator fluorescent dyes and/or different DNAs or oligonucleotides of different base pair combinations should yield better detection limits and improved selectivity. Further understanding of the relationship between intercalation and carcinogens may allow the development of systems for rapid sensitive detection of carcinogens in environmental applications.

Acknowledgement

We would like to thank the Environmental Protection Agency, Inter-agency agreement #DW13934923-01-2, for their support of this work.

Literature Cited

1. Michaelis, Cold Spring Harbor Symposia on Quantitative Biology, **XII** ,72,(1947)
2. Peacock, R.A., and Skerrett, J.N.H., *Trans. Faraday Soc.*, **1956**, 52, p.261.
3. Bresloff, J.L., and Crothers, D.M., *Biochemistry*, **1981**, 20 pp. 3547-3553.
4. Scatchard, G., Ann. *N.Y. Acad. Sci.*, **1949**, p. 600.
5. Cavalier, L.F., Rosoff, M., and Rosenberg, J. *Amer. Chem. Soc.*, **1956**, 78, p. 5239.
6. Wilson, W.D., and Jones, R.L., *in Intercalation Chemistry*, edit. Whittingham, M.S. and A.J. Jacobson, Academic Press, New York, **1982**, pp. 445-501.
7. Richardson, C.L., and Schulman, G.E., *Biochimica et Biophysica Acta.*, **1981**, 652, pp. 55-63.
8. Shahbaz, M., Harvey, R.G., Prakash, A.S., Boal, T.R., Zegar, I.S., and LeBreton, P.R., *Biochem. Biophys. Res Commumn.*, **1983**,m 112, pp. 1-7.
9. Zegar, I.S., Prakash, A.S., and LeBreton, P.R., J.*Biomolecular Structure and Dynamics*, **1984**, 2, pp. 531-542.
10. LeBreton, P.R., *Amer. Chem. Soc.*, **1985**, 9, pp.209-238.
11. Dinesen, J., Jacobson, J.P., Hansen, F.P., Pedersen, E.B., and Eggert, H., J. *Med. Chem.*, **1990**, 33, p. 93.
12. Nordmeier, E., J. *Phys. Chem.*, **1992**, 96, pp. 6045-6055.
13. Neidle, N., Pearl, L.H., Herzyk, P., and Berman, H.M., *Nucleic Acids Research*, **1989**, 16, pp. 8999-9016.
14. Zimmerman, S.C., Lamberson, C.R., Cory M., and Fairley, T.A., J. *Amer. Chem. Soc.*, **1989**, 111, p. 6805.
15. Tanious, F.A., Veal, J.M. Buczak, H., Ratmeyer, L.S., and Wilson, W.D., *Biochem.*, **1992**, 31, pp. 3103-3112, 1992.
16. Lerman, L.S., *Proc. Natl. Acad. Sci., U.S.A.*, **1962**, 49, pp. 94-101.
17. Lerman, L.S., *Cell. Comp. Physiol (Suppl. 1)*, **1964**, 64, pp. 39-48.
18. Kapuscinski, J., and Darzynkiewics, Z., J.*Biomolecular Structure and Dyanmics*, **1987**, 5, pp. 127-143.
19. McFadyen, W.D., Sotirellis, N., Denny, W.A., and Waklin, L.P.G., *Biochem. Biophys. Acta*, **1990**, 1048, p. 50.
20. Wilson, W.D., Tanious, F.A., Barton, H.J., Strekowski, L., Boykin, D.W., and Jones, R.L., J.*Amer. Chem. Soc.*, **1989**, 111, p. 5008.
21. Harvey, R.G., and Geacintov, N.E., *Acc. Chem. Res.*, **1988**, 21, p. 66.
22. Kim, S.K., Geacintov, N.E., Brenner, H.C., and Harvey, R.G., *Carcinogenisis*, **1989**, 10, p. 1333.
23. Armstrong, J.A., and Niven, J.S.F., Nature, **1957**, 180, p. 1335.
24. United States patent #4,257,774; March 24, 1981 - *"Intercalation inhibition assay for compounds that interact with DNA or RNA,"* Carol L. Richardson and Gail E. Schulman.
25. Gibco BRL, *Catalogue and Reference Guide*, Life Technologies, Inc., Gaithersburg, MD, **1991.**
26. Lakowicz, J.R., *Principles of Fluorescence Spectroscopy*, Plenum Press, New York, **1984**

27. Schelhorn,T., Kretz,S., and Zimmermann,H.W., *Cell. Molec. Biology*, **1992**, 38, pp. 345-365.
28. Neidle,S., and Abraham, Z., *CRC Critical Reviews in Biochemistry*, **1984**, 17, pp. 73-121.
29. Herzyk, P., Neidle, S., and Goodfellow, J., *Biomolecular Structure Dyanmics*, **1992**, 10, pp. 97-139.
30. Lesko, S.A., Smith, A., Ts'o, P.O.P., and Umans, R.S., *Biochemistry*, **1968**, 7, pp. 434-437.
31. Craig, M., and Isenberg, I., *Biopolymers*, **1970**, 9, pp. 689-696.
32. Isenberg, I., Baird, S.L., and Bersohn, R., Ann. *N.Y. Acad. Sci,* **1969**, 153, p. 780.
33. Isenberg, I., Baird, S.L., and Bersohn, R., *Biopolymers*, **1967**, 5, p.477.
34. Nelson, H.P., and DeVoe, H., *Biopolymers*, **1984**, 23, pp. 897-911.
35. Hermann, R.B., J. *Phys. Chem.*, **1972**, 76, pp. 2754-2759.
36. Caillet, J., and Pullman, B., *Molecular Association in Biotechnology*, Pullman, B., Ed., Academic Press, New York, **1968**, pp. 217-220.
37. Haugland, R.P., Molecular Probes "*Handbook of fluorescent probes and research chemicals* (Larions K.D. Edit). **1992.**
38. National Academy of Sciences, *Particulate Polycyclic Organic Matter*, National Academy of Sciences, Washington, DC **1972.**
39. Wilson, W.D., Tanious, F.A., Barton, H.J., Jones, R.L., Fox, K., Wydra, R.L., and Strekowski, L., *Biochemistry*, **1990**, 29, pp. 8452-8461.
40. Larsen, T.A., Goodsell, D.S., Cascio, D., Grzeskwaik, K., and Dickerson, R.E., J. *Biomolecular Structure and Dynamics*, **1989**, 7, pp. 477-491.
41. Menzie, C.A, Potocki, B.B, Santodonato, J., *Environ. Sci. Technol., **1992**,* 26, pp. 1278-1284.

RECEIVED August 28, 1995

Chapter 6

Chemically Modified Electrode for Hydrogen Peroxide Measurement by Reduction at Low Potential

Ashok Mulchandani and Lisa C. Barrows

Chemical Engineering Department, College of Engineering, University of California, Riverside, CA 92521

This paper reports the development of a novel chemically modified electrode for cathodic determination of hydrogen peroxide. The electrode was constructed by modifying the surface of a glassy carbon electrode with an electrochemically deposited ferrocene-modified polyaniline film from a solution of N-(ferrocenylmethyl)aniline monomer in acetonitrile. Hydrodynamic voltammetry studies showed that both hydrogen peroxide and oxygen were reduced at the electrode with an increasing response at higher cathodic potentials. The interference due to molecular oxygen was minimized at -50 mV vs. Ag/AgCl. The response of the electrode was a function of the thickness of the poly(anilinomethylferrocene) film and pH of the electrolyte, and was free from interference due to electroactive compounds such as ascorbic acid, uric acid, and acetaminophen. Application of the new chemically modified electrode for the construction of a flavin enzyme (glucose oxidase)-based biosensor is demonstrated.

Enzyme electrodes based on detection of hydrogen peroxide, the product of flavin oxidase enzymes, are the most widely used of all enzyme electrodes. In these electrodes, the hydrogen peroxide formed is detected amperometrically using a platinum working electrode poised at 0.6-0.7 V vs. Ag/AgCl reference (1). At such high potential, compounds such as ascorbic acid, uric acid, and amino acids, which are common components of biological fluids, are also oxidized. Hence, the selectivity of such enzyme electrodes is low. To overcome the limitations imposed by the requirement of large overpotentials on the selectivity of these sensors, different approaches have been described. In a commercially available glucose and lactate analyzer (Yellow Springs Instrument, Yellow Springs, OH) three relatively thick membranes, one of which is permeable to H_2O_2 only, are used to alleviate the interference problem (2). This results in a high diffusion resistance for the reactant reaching the enzyme and H_2O_2 reaching the electrode and hence affects the response time. Platinized and rhodinised carbon electrodes have been developed to lower the oxidation potential for H_2O_2 to +0.4 V (3). Although the oxidation potential for H_2O_2 on such electrodes is lowered, detection is still not completely

0097–6156/95/0613–0061$12.00/0

free of interference (4). Determination of H_2O_2 by reduction at electrodes modified with $Ru(NH_3)_6^{3+}$ incorporating montmorillonite clay (5), palladium/iridium (6), and iron phthalocyanine (7) have been reported. These electrodes, however, have limitations. The determination of H_2O_2 at the Pd/Ir modified electrode is performed at -0.3 V (vs. Ag/AgCl), a potential at which dissolved oxygen is also reduced and therefore interferes. The reduction of H_2O_2 at $Ru(NH_3)_6^{3+}$ incorporated montmorillonite clay and iron phthalocyanine modified electrodes require strongly acidic environment (pH 2) and therefore limit the application of these electrodes for construction of flavin enzyme-based biosensors. Use of enzyme (peroxidase) modified electrodes to determine H_2O_2 by reduction has been reported. These enzyme modified electrodes, operating between 0 and -0.2 V vs. Ag/AgCl, are reported to be free of interference from ascorbate, urate, and paracetamol (8-13).

In this paper, the development and characterization of a poly(anilinomethyl-ferrocene) modified glassy carbon electrode for determination of hydrogen peroxide by reduction at low applied potential is reported.

EXPERIMENTAL SECTION

Chemicals. (2-[N-Morpholino]ethanesulfonic acid) monohydrate (MES) and glucose oxidase (EC 1.1.3.4) type VII from *Aspergillus niger* (activity 1682 U/g solid) were purchased from Sigma Chemical Co. (St. Louis, MO). Sodium phosphate monobasic monohydrate, sodium phosphate dibasic anhydrous, citric acid monohydrate, acetonitrile (HPLC grade), tetrabutylammonium perchlorate (TBAP), glutaraldehyde (25% in water), and hydrogen peroxide (30% in water) were obtained from Fisher Scientific (Tustin, CA). Aniline, 1,2 phenylenediamine, and resorcinol were acquired from Aldrich (Milwaukee, WI). All the chemicals were used without purification. Double distilled ultrapure water was used for preparation of the buffers, standards, and electrochemistry work.

Synthesis of N-(ferrocenylmethyl)aniline monomer (I). Monomer **I** was synthesized according to the procedure reported earlier (14).

Electrode construction. Glassy carbon electrodes (GCEs) (Bioanalytical Systems Inc., Lafayette, IN) were polished with 1 mm diamond paste followed by 0.05 mm g-alumina particles (Buehler, Lake Bluff, IL). Electrodes were rinsed with water and ultrasonicated for 2-5 min after each polishing step. A poly(anilinomethylferrocene), designated as poly(AMFc), film was deposited on the prepared GCE from an electrolyte bath containing 5 mM **I** and 0.1 M TBAP in acetonitrile, which was deaerated with argon prior to electropolymerization, by cycling the electrode potential at 100 mV s^{-1} between 0 and 1.1 V vs. Ag/AgCl for the desired number of cycles. The electrode was then rinsed with acetonitrile to remove entrapped monomer and TBAP and stored dry.

Polyaniline modified GCEs were prepared by depositing polyaniline film on polished GCE from 0.1 M aniline in 1.0 M perchloric acid and 0.1 M aniline plus 0.2 M TBAP in acetonitrile by cycling the potential at 100 mV s^{-1} between -0.2 and 0.9 V vs. Ag/AgCl for 5 cycles.

Glucose enzyme electrodes were constructed by immobilizing glucose oxidase on the top of the poly(AMFc) film by crosslinking with glutaraldehyde and subsequent entrapment in a 1,2-phenylenediamine - resorcinol polymer film. Three μl of a mixture containing 45 μg of glucose oxidase enzyme and 1% (w/v) glutaraldehyde in 0.1 M pH 6.5 phosphate buffer was added on the top of the poly(AMFc) film and allowed to dry for 2 h at room temperature. Subsequently, this electrode was made the working electrode in an electrochemical cell containing a deaerated solution of 1.5 mM each of resorcinol and 1,2-phenylenediamine monomers in 0.1 M pH 6.5 phosphate buffer with 0.1 M $NaClO_4$. The potential was cycled between 0 and + 0.65 V vs. Ag/AgCl at a scan rate of 20 mV/s for 8 cycles to deposit a polymer film. The electrode was washed in 0.1 M pH 5.5 citrate-phosphate to remove any free enzyme and monomers.

Measurements. Electropolymerizations were performed under stationary conditions in a 10 ml electrochemical cell placed inside a Faraday cage (BAS, C-2), with a potentiostat/galvanostat (263A, EG&G, Princeton, NJ) interfaced to an 80486-based personal computer. Batch amperometric measurements were performed under stirred conditions in a 10 ml electrochemical cell placed inside a Faraday cage (BAS, C-2), with a Voltammograph (BAS, CV 27) coupled to a low current module (BAS, PA1-Preamplifier). The signals were recorded on an X-Y-t chart recorder (BAS, MF-8051). All measurements were performed with Ag/AgCl reference and platinum auxiliary electrodes in either 0.1M citrate-phosphate buffer with 0.1 M $NaClO_4$, 0.05 mM pH 5.5 MES buffer with 0.1 M $NaClO_4$, pure acetonitrile with 0.1 M TBAP, or 90% acetonitrile plus 10% 0.05 M pH 5.5 MES buffer with 0.1 M TBAP.

RESULTS AND DISCUSSION

Hydrodynamic voltammetry. Figure 1 shows the dependence of the poly(AMFc) modified electrode response on the applied potential to additions of deaerated H_2O_2 solution and non-deaerated water to deaerated 0.05 M pH 5.5 MES buffer with 0.1 M $NaClO_4$ in the electrochemical cell. In presence of H_2O_2, while there was no response between +0.3 and +0.2 V, the response increased rapidly as the potential was made more cathodic to -0.2 V. When evaluated for response to non-deaerated water over the same potential range, the poly(AMFc) electrode had no response between potentials of +0.3 and -0.05 V and a small cathodic response at potentials of -0.1 and -0.2 V. The response at -0.1 and -0.2 V is attributed to the reduction of molecular oxygen at the poly(AMFc) modified electrode. The potential of -0.05 V, at which there was no significant interference due to oxygen reduction, was therefore selected as the operating potential in subsequent work.

Effect of number of electrochemical polymerization scans. Figure 2 shows the plot of the cathodic response of the poly(AMFc) modified electrode to H_2O_2 as a function of the number of polymerization scans used for depositing the polymer film. As expected, due to the higher ferrocene surface coverage, the response initially increased with an increase in number of polymerization scans. However, on further increase in number of polymerization scans, the response decreased. This decrease is presumably associated with charge transfer limitations arising due to a thicker polymer film deposited on the GCE with increasing numbers of polymerization scans. No attempts were made to determine the thickness of deposited films. Electrodes constructed using 5 polymerization scans were used in subsequent studies.

Effect of pH. The cathodic response of the poly(AMFc) modified electrode to H_2O_2 increased as the pH of the measuring electrolyte was lowered (Fig. 3A). This inverse relationship between the cathodic response and the electrolyte pH can be attributed to the increase in the conductivity of the polyaniline backbone of poly(AMFc) (15,16). The ability of the poly(AMFc) modified electrode to monitor

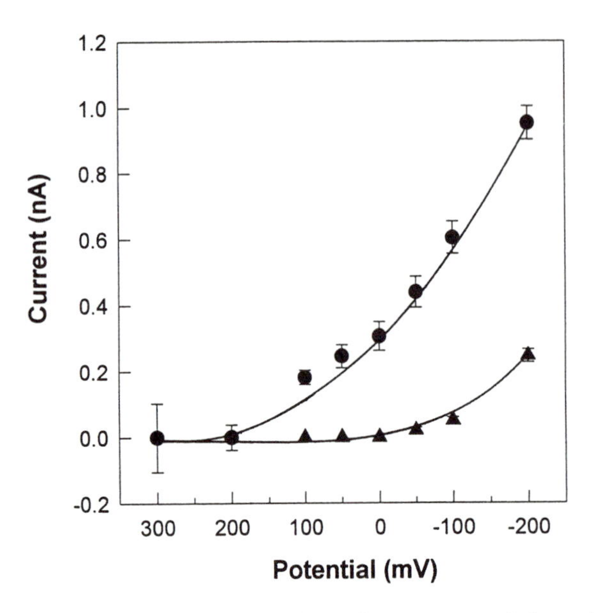

Figure 1. Effect of operating potential on the response of the poly(AMFc) modified electrode to (•) 10 µM H_2O_2 and (▲) 10 µL injections of water in deaerated (10 mL) 0.05 M MES pH 5.5 buffer with 0.1 M $NaClO_4$.

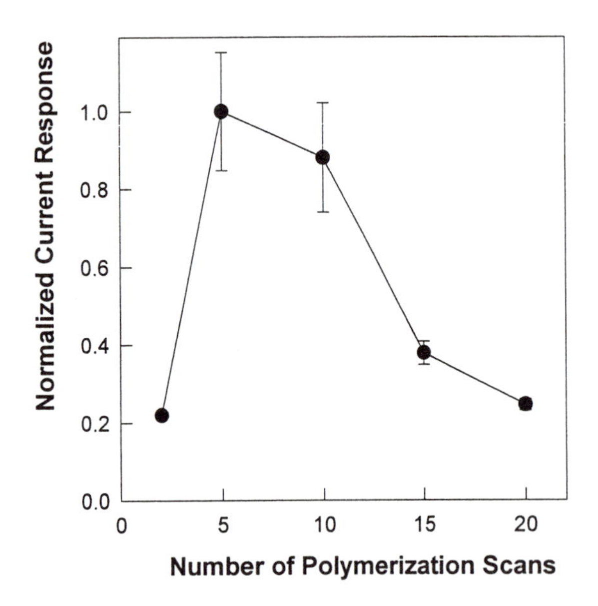

Figure 2. Effect of number of polymerization scans on the cathodic response of the poly(AMFc) modified electrode to 10 µM H_2O_2 in pH 5.5 0.1 M MES buffer with 0.1 M $NaClO_4$ at -50 mV vs. Ag/AgCl reference.

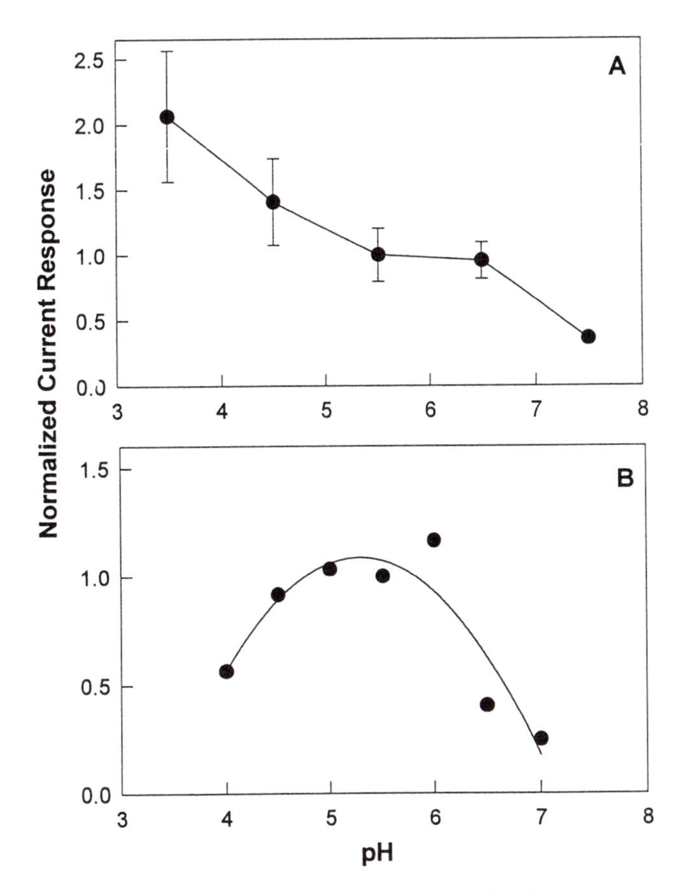

Figure 3. Effect of the solution pH on the cathodic response of the poly(AMFc) modified electrode to (A) 10 μM H_2O_2 in 0.15 M $NaClO_4$ solution adjusted to the desired pH by HCl or NaOH and (B) 100 μM β-D glucose in 0.1 M citrate-phosphate buffer with 0.1 M $NaClO_4$. Operating potential -50 mV vs. Ag/AgCl reference.

H_2O_2 over a wide pH range demonstrates its applicability in construction of oxidase enzyme-based amperometric biosensors.

Control experiments were performed to investigate if GCEs modified with polyaniline alone responded to H_2O_2 at the selected potential (-50 mV) and pH (5.5). Electrodes modified with polyaniline films grown from aniline solution in acetonitrile and pH 1.0 aqueous medium did not respond (data not shown). These results are in agreement with the data of Doubova et al. (17), who report that, even at low pH, the cathodic reduction of H_2O_2 at the polyaniline modified electrode is extremely inhibited because of the very low kinetic efficiency of H_2O_2 for oxidizing aniline. The fact that the polyaniline modified electrode does not detect H_2O_2 by reduction, whereas the poly(AMFc) modified electrode does, leads us to hypothesize that the ferrocene incorporated in the polyaniline film is oxidized by the peroxide to ferricinium, which is then electroreduced at the electrode surface. This hypothesis is also based on literature reports that H_2O_2 can be determined by the oxidation of (a) Fe(II) to Fe(III) by peroxides at low pH (2.0), followed by complexation with xylenol orange and determination of the complex spectrophotometrically (18) and (b) by the oxidation of Fe(II)PC (iron phthalocyanine) incorporated in carbon paste to Fe(III)PC at pH 2.0 and subsequent electroreduction of Fe(III)PC at +0.1 to 0.2 V vs. Ag/AgCl (7).

Analytical characterization. Figure 4 shows the calibration plot of the poly(AMFc) electrode operating at -50 mV vs. Ag/AgCl in aqueous medium (0.05 M pH 5.5 MES buffer with 0.1 M $NaClO_4$). The sensitivity of the electrode in the linear range (0 - 30 μM) was 0.167 nA μM^{-1}. The precision of the electrode expressed in terms of the relative standard deviation was 3% for a concentration level of 25 μM H_2O_2 (n=25). When used repeatedly for 100 analyses (25 μM H_2O_2) over a period of 2.5 h in a flow injection analyzer, the output response of the electrode gradually dropped to approximately 67% of the original response. When compared to the peroxide electrode constructed using the poly(AMFc) and horseradish peroxidase, the sensitivity of the CME developed in this work was 50% lower (14).

In many applications the analyte to be monitored is present in an organic matrix that is either not soluble or has very low solubility in aqueous media and therefore monitoring in organic media is preferable. The recent remarkable findings that enzymes can function in organic media has attracted considerable interest in developing organic phase enzyme electrodes (19). Experiments were therefore performed in order to evaluate the applicability of the poly(AMFc) electrode to monitor H_2O_2 in organic solvents. The electrode was able to detect H_2O_2 in pure acetonitrile and in a 90% acetonitrile and 10% 50 mM pH 5.5 MES buffer with 0.1 M TBAP mixture. Since the response of the electrode to the same concentration of hydrogen peroxide (10 μM) was 400% higher in the 10% aqueous and 90% organic mixture compared with the pure organic solvent, experiments for electrode calibration plots (Fig. 4) were conducted in the former mixture. The sensitivity of the electrode to H_2O_2 in the linear range (0 - 20 μM) was 0.535 nA μM^{-1}. As in the aqueous phase, the sensitivity of the poly(AMFc) modified electrode in predominantly organic medium was lower than that of the electrode incorporating both poly(AMFc) and horseradish peroxidase together (19).

Interference studies. In the conventional electrochemical method for H_2O_2 determination based on the oxidation at either platinum or carbon electrodes, electroactive compounds such as ascorbic acid, acetaminophen, and uric acid interfere. Various approaches, such as the incorporation of peroxidases or

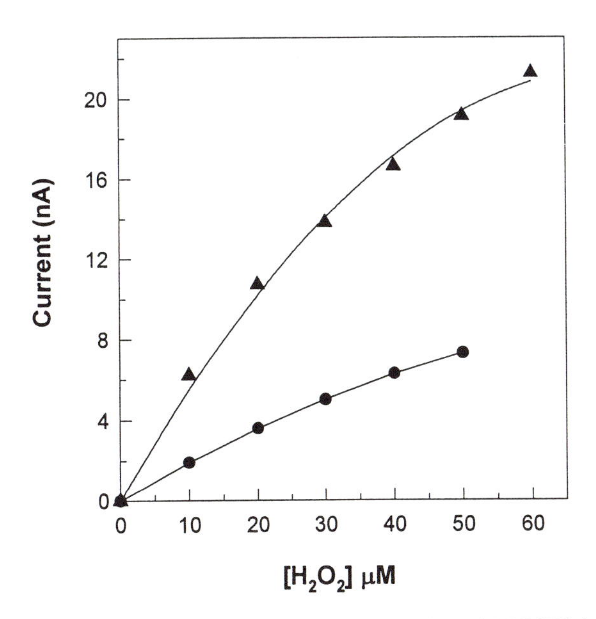

Figure 4. Calibration plots for (•) H_2O_2 in pH 5.5 0.05 M MES buffer with 0.1 M $NaClO_4$ and (▲) H_2O_2 in 90% acetonitrile with 0.1 M TBAP plus 10% 0.05 M pH 5.5 MES buffer with 0.1 M $NaClO_4$. Operating potential -50 mV vs. Ag/AgCl. Data points are average of measurements using two electrodes.

membranes, have been tried to prevent unwanted reactions from taking place at the electrode surface (20). Another approach to prevent undesired reactions from taking place at the electrode surface is to perform the amperometric detection around -100 to 0 mV vs. SCE (20). When the poly(AMFc) modified electrode was operated at pH 5.5 and at applied potential of -50 mV vs. Ag/AgCl, the response to H_2O_2 was unaffected by the presence of 0.06 mM ascorbic acid, 2 mM acetaminophen and 0.03 mM uric acid.

Application of poly(AMFc) electrode. The poly(AMFc) modified electrode was used in the construction of a flavin enzyme-based biosensor. Fig. 3B shows the effect of pH on the response of the glucose electrode. This pH profile is similar to that reported in literature with a maximum at pH of around 5.5. A pH of 5.5 was therefore used for characterization of the electrode.

The calibration plot and characteristics of a glucose oxidase modified poly(AMFc) electrode for the determination of β D-glucose is shown in Fig. 5. The response of the glucose biosensor was linear between 0 and 190 μM and the sensitivity in the linear range was 0.0145 nA μM^{-1}. Investigations into the response of the enzyme electrode to glucose in oxygen free medium showed that the electrode produced no response, confirming that the electrode response was only due to the H_2O_2 produced by the enzyme catalyzed reaction of glucose and oxygen. Similar to the H_2O_2 measurement, glucose measurements were not affected by the presence of 0.06 mM ascorbic acid and 0.03 mM uric acid.

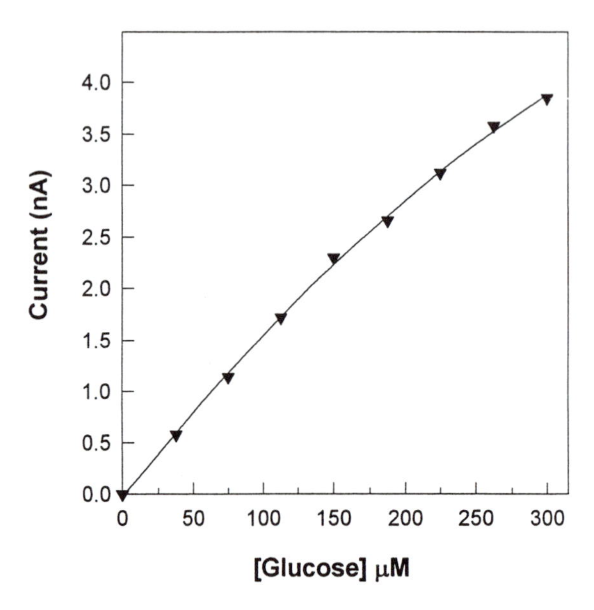

Figure 5. Calibration plots for β-D glucose in pH 5.5, 0.05 M citrate-phosphate buffer with 0.1 M $NaClO_4$. Operating potential -50 mV vs. Ag/AgCl. Data points are average of measurements using two electrodes.

CONCLUSIONS

The use of poly(anilinomethylferrocene) modified electrode as the basis of hydrogen peroxide detection by reduction has been demonstrated in both aqueous and organic media. Operating at a low potential of -50 mV vs. Ag/AgCl, the response of the sensor was not interfered by molecular oxygen, ascorbic acid, acetaminophen, and uric acid. When modified with glucose oxidase, the biosensor was able to detect glucose, thereby demonstrating the application of the poly(AMFc) modified electrode in the construction of flavin enzyme-based biosensors.

ACKNOWLEDGMENT

This work was financially supported by the National Science Foundation Grant BCS 9309741.

LITERATURE CITED

1. Guilbault, G.G.; Lubarano, G.J. *Anal. Chim. Acta* **1973**, *64*, 439-455.
2. Kulys, J.J.; Laurinavicius; V.-S.A.; Pesliakiene M.V.; Gureviciene, V.V. *Anal. Chim. Acta* **1983**, *148*, 13-18.
3. White, S.F.; Turner, A.P.F.; Schmid, R.D.; Bilitewski, U.; Bradley, J. *Electroanalysis* **1994**, *6*, 625-632.
4. Hajizadeh, K.; Halsall, H.B.; Heineman, W.R. *Anal. Chim. Acta* **1991**, *243*, 23-32.

5. Oyama, N.; Anson, F.C. *J. Electroanal. Chem.* **1986**, *199*, 467-470.
6. Cox, J.A.; Jaworski, R.K. *Anal. Chem.* **1989**, *61*, 2176-2178.
7. Qi, X.; Baldwin, R.P. *Electroanalysis* **1993**, *5*, 547-554.
8. Kulys, J.J.; Pesliakiene, M.V.; Samalius, A.S. *Bioelectrochem. Bioenerg.* **1981**, *8*, 81-88.
9. Wang, J.; Freiha, B.; Naer, N.; Romero, E. G.; Wollenberger, U.; Ozsoz, M. *Anal. Chim. Acta* **1991**, *254*, 81-88.
10. Vreeke, M.; Maidan, R.; Heller, A. *Anal. Chem.* **1992**, *64*, 3084-3090.
11. Wollenberger, U.; Bogdanovskaya, V.; Borbin, S.; Scheller, F.; Tarasevich, M. *Anal. Lett.* **1990**, *23*, 1795-1808.
12. Gorton, L.; Jönsson-Pettersson, G.; Csöregi, E.; Johansson, J.; Domìnguez, E.; Marko-Varga, G. *Analyst* **1992**, *117*, 1235-1241.
13. Mori, H.; Kogure, M.; Kukambe, K. *Anal. Lett.* **1992**, *25*, 1643-3656.
14. Mulchandani, A.; Wang, C-L.; Weetall, H.H. *Anal. Chem.* **1995**, *67*, 94-100.
15. Horwitz, C.P.; Dailey, G.C. *Chem. Mater.* **1990**, *2*, 343-346.
16. Deshpande, M.V.; Amalnerkar, D.P. *Prog. Polym. Sci.* **1993**, *18*, 623-649.
17. Doubova, L.; Mengoli, G.; Musiani, M.M.; Valcher, S. *Electrochim. Acta* **1989**, *34*, 337-343.
18. Gupta, B.L. *Microchem. J.* **1973**, *18*, 363-374.
19. Wang, C-L.; Mulchandani, A. *Anal. Chem.* **1995**,*67*, 1109-1114.
20. Gorton, L.; Bremle, G.; Csöregi, E.; Jönsson-Petersson, G.; Persson, B. *Anal. Chim. Acta* **1991**, *249*, 43-54.

RECEIVED June 30, 1995

Chapter 7

Enzyme Sensors for Subnanomolar Concentrations

Frieder W. Scheller[1], Alexander Makower[1], Andrey L. Ghindilis[2], Frank F. Bier[1], Eva Förster[1], Ulla Wollenberger[1], Christian Bauer[1,7], Burkhard Micheel[3], Dorothea Pfeiffer[4], Jan Szeponik[4], Norbert Michael[5], and H. Kaden[6]

[1]Analytical Biochemistry, Institute of Biochemistry and Molecular Physiology, University of Potsdam, c./o. Max-Delbrück-Center of Molecular Medicine, Robert-Rössle-Strasse 10, D–13122 Berlin, Germany
[2]Research Center of Molecular Diagnostics and Therapy, Department of Biosensors, Simpheropolsky blvd. 8, 113149 Moscow, Russia
[3]Max-Delbrück-Center of Molecular Medicine, Robert-Rössle-Strasse 10, D–13122 Berlin, Germany
[4]BST Bio Sensor Technology GmbH Berlin, Buchholzerstrasse 55–61, D–13156 Berlin, Germany
[5]Research Institute of Molecular Pharmacology, Alfred-Kowalke-Strasse 4, D–10315 Berlin, Germany
[6]Kurt-Schwabe-Institute for Measuring and Sensor Technology, Fabriskstrasse 69, D–04736 Meinsberg, Germany

Similar to the principles of biochemical signal amplification, cyclic reaction can provide an effective increase of sensitivity both in amperometric enzyme electrodes and enzyme immunoassays.
In this study we used the copper enzyme laccase (EC1.10.3.2) from *Coriolus hirsutus,* which catalyzes the oxidation of a wide range of substances (among them epinephrine) by dissolved oxygen. To complete the cycle the reduction of the formed quinones or ferric compounds is accomplished by pyrroloquinoline quinone (PQQ) containing (NADH independent) glucose dehydrogenase (EC1.1.99.17), which converts ß-D-glucose to gluconolactone.
Owing to the broad spectrum of substrates for both enzymes, the sensor responds to various catecholamines, aminophenols and ferrocene derivatives. The best sensitivities were obtained for aminophenol and epinephrine where the lower limit of detection is 100 pM. The recycling sensor was used to trace the secretion of catecholamines in cultures of adrenal chromaffin cells. Furthermore both a sandwich assay for IgG and displacement of enzyme labeled cocaine have been indicated by the amplification enzyme sensor.

In the liver cell cascade-like sequential activation of enzymes provides a tremendous signal amplification of more than five orders of magnitude. This process is initiated by the binding of

[7]Corresponding author

0097–6156/95/0613–0070$12.00/0

a hormone, epinephrine, to the relevant receptor and contributes to the regulation of the blood glucose level. During the following glycolysis small concentration changes are amplified by the cyclic conversion of the key metabolites (1). Here we report an enzyme recycling system where epinephrine, at the nanomolar level, triggers the consumption of an amount of glucose that is higher by four orders of magnitude.

The high sensitivity of these systems results from the cyclic conversion of a "shuttle" molecule by two enzymes. In each cycle one diffusible species is formed, which transfers the chemical signal to a transducer. In analogy to metabolic cycles the shuttle molecule flips between its reduced and oxidized or phosphorylated and dephosphorylated state (2). The cyclic reactions are catalyzed by appropriate pairs of enzymes, such as oxidases/dehydrogenases. The shuttle can be any analyte for which such a pair can be constructed, e.g. lactate glutamate, phosphate, ATP or NAD(P)H (3-11).

Experimental

Laccase (Lacc) (300 - 500 U mg^{-1}) from *Coriolus hirsutus* was purified and supplied by the Research Center for Molecular Diagnostics, Moscow (Russia). Quinoprotein glucose dehydrogenase (GDH) (950 U mg^{-1}) from *Acinetobacter calcoaceticus,* p-aminophenol (PAP) and p-aminophenol phosphate (PAPP) was kindly supplied by Boehringer Mannheim. All chemicals were used without further purification.

For preparation of the bienzyme membrane a suspension of 400 mg polyvinyl alcohol (Sigma) in 2 ml H_20 was incubated for 5 h and then heated to 100° C for 5 min. to form a solution. After cooling, the polyvinyl alcohol solution was mixed with an equal volume of Lacc (9000 U ml^{-1}) and GDH (9000 U ml^{-1}) solution and applied to the surface of 4 mm diameter plastic disks (2 µl on each disk). The coated disks were irradiated by UV light for 30 min. to facilitate polymerization. The membranes were removed from each disk and stored at 4° C.

For measurements an enzyme membrane was placed on the surface of an oxygen electrode (diameter 0.5 mm), covered by a dialysis membrane and fixed by an O-ring. Measurements were performed in a 1.2 ml stirred cell or in a flow cell.

The Laccase/Glucosedehydrogenase Sensor

The principle of enzymatic signal amplification for PAP determination is illustrated in Fig. 1 Lacc catalyzes the oxidation of PAP to p-iminoquinone under oxygen consumption. The latter is the electron acceptor for the glucose oxidation catalyzed by GDH which thus regenerates PAP. In this way an excess of glucose enables the detection of low PAP concentration by indication of the increased oxygen depletion.

Figure 2. shows the electrode response upon PAP at a concentration of 25 mM glucose and in the absence of glucose. In the presence of glucose a detection limit of 100 pM PAP is reached. The detection limit without glucose is 500 nM PAP, i.e. at a 5000 times higher concentration.

Lacc has a broad substrate specificity, it oxidizes a variety of diphenols, inorganic and organic redox dyes, amino substituted phenols and catecholamines (12). On the other hand GDH accepts a number of redox mediators for glucose oxidation. Therefore, any substrate of Lacc, which oxidized form is accepted by GDH can be determined by the Lacc-GDH electrode. Consequently, the sensor response is not selective for a certain substrate. However, this is not a problem for the cases described below, when only one of the substrates is present (immunoassays) or a mixture with unchanging ratio of catecholamines (norepinephrine and epinephrine) secretion is analyzed. The dependence of electrode response on the concentration of 8 different substrates is presented in Fig. 3. The optimal substrate is PAP, followed by epinephrine, ferroceneacetic acid, ferrocenecarboxylic acid, L-DOPA, arterenol, 1,1'-ferrocendicarboxylic acid, and norepinephrine. It should be noted that for ascorbic acid no amplification was observed.

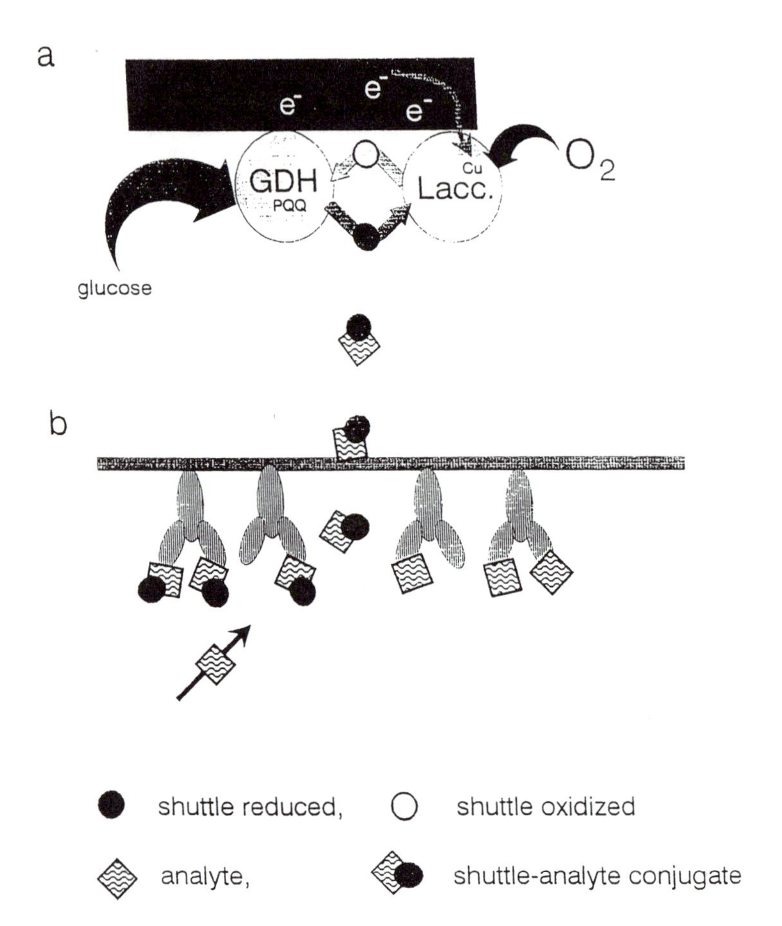

Figure 1. (a) Scheme of the amplification cycle. The shuttle molecule is cycled between the two enzymes glucose dehydrogenase (GDH) and laccase stimulating the consumption of O_2 and glucose. (b) Such a cycle can be connected to an immunoassay using a shuttle-hapten conjugate, which is released by the binding of an analyte-hapten. The cycle works as a "biochemical transistor" if the laccase is connected to an electrode facilitating direct electron transfer as indicated at the top of the scheme.

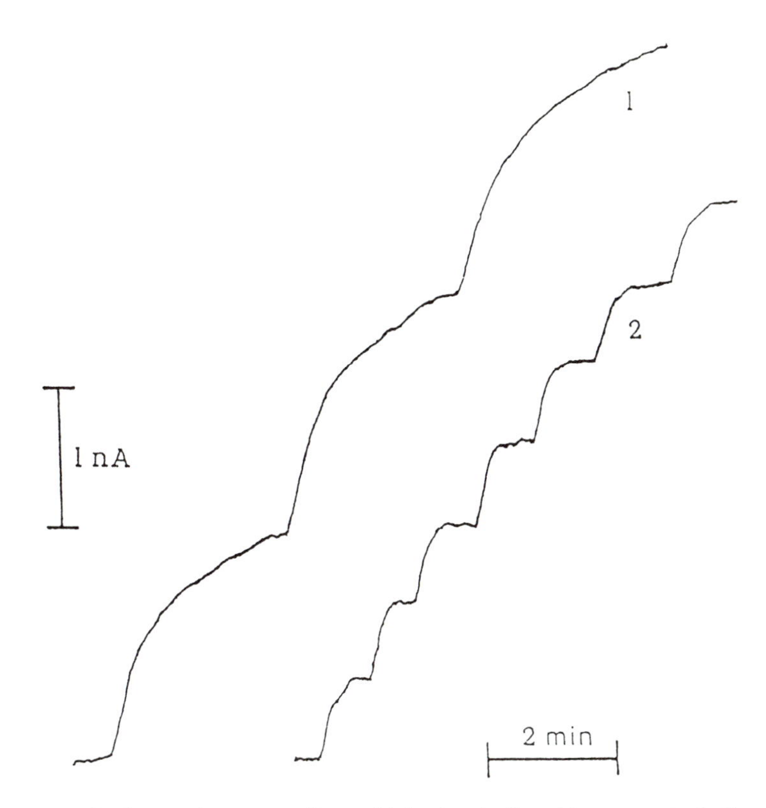

Figure 2. Current-time recording obtained at the bienzyme electrode on increasing the p-aminophenol (PAP) concentration in 1 nM (1) and 10 μM (2) steps. 1- in the presence of glucose (25 mM); 2 - in glucose-free solution.

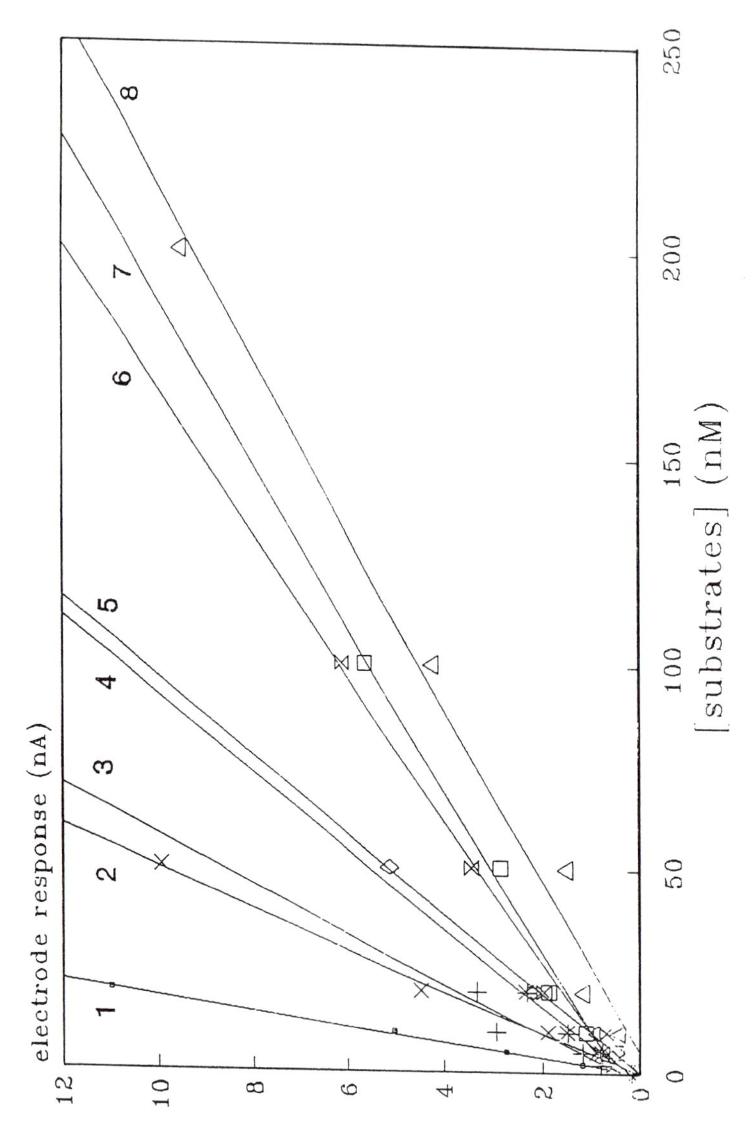

Figure 3. The concentration dependence of electrode response: PAP (1), epinephrine (2), ferroceneacetic acid (3), ferrocenecarboxilic acid (4), L-DOPA (5), arterenol, (6), 1-1 ferrocenedicarboxilic acid (7), norepinephrine (8). Conditions: 0.1 M phosphate buffer pH 6 containing 25mM of glucose.

The cyclic reaction is driven by the oxidation of glucose by oxygen. The comparison of the sensitivities in the absence and presence of glucose indicates that for each epinephrine molecule more than five thousand glucose molecules are converted, i.e. the addition of 10 nM hormone leads to a consumption of 50 µM glucose. In this respect the action of epinephrine resembles that of the hormone insulin towards the fat cell.

The enormous efficiency of the *amplification sensor* is based on the great excess of enzyme molecules (100 µM) compared with the concentration of the shuttle molecule (100 pM -10 nM) within the reaction layer. Therefore the internal transport of the recycled substance between the two enzymes is very fast. The current density of this membrane-covered sensor is almost three orders of magnitude higher than at a bare electrode (Figure 4). In this respect the enzyme cycle represents a chemical signal amplifier coupled to the transducer by oxygen diffusion. By analogy to the recently described diode-like electrochemical behavior of succinate dehydrogenase (13), direct electron transfer between a recycling enzyme and a redox electrode would represent a "biochemical transistor".

Direct communication between the active site of Lacc and an electrode surface has been reported (14). Using carbon electrodes, mediatorless catalysis of oxygen reduction near the equilibrium potential of the four electron reduction was established. We found that addition of a phenolic substrate leads to a potential shift from the equilibrium of the Lacc modified electrode in the cathodic direction due to the consumption of oxygen. By sequential fixation of Lacc and GDH to a polyethylenimine modified carbon fiber we coupled analyte recycling and mediatorless electron transfer. In this sensor Lacc fulfills two functions (i) it oxidizes the reduced form of the analyte, e.g. epinephrine or aminophenol, (ii) it catalyzes the mediatorless electron transfer from the remaining oxygen to the carbon electrode. Therefore the addition of the reduced substrate shifts the electrode potential to a more cathodic steady state value which is determined by the decreased oxygen concentration. In the presence of glucose considerably more oxygen is consumed by the cyclic analyte conversion (as is illustrated in Figure1). For epinephrine the limit of detection is 20 nM. Obviously, the lower enzyme loading compared with the membrane type sensor described above leads to a loss in sensitivity.

Direct combination of biochemical and electronic signal amplification has been achieved by immobilizing the enzyme couple Lacc / GDH on the gate of a pH-ISFET.

As has been already mentioned in the bienzymatic recycling process quinone species formed during oxidation of biphenols by Lacc are reduced by GDH in presence of glucose. In the GDH reaction glucose is oxidized to gluconolactone which hydrolyses to gluconic acid leading to a decrease in pH. For PAP as well as for the neurotransmitter norepinephrine the lower detection limit is about 10 nM. Below 100 nM sensitivity for PAP and norepinephrine was 0.8 and 0.5 mV/decade respectively For higher concentrations up to 1000 nM sensitivity increased to 1.6 - 2.0 mV/decade (Figure 5). The small signals are caused by the slow spontaneous hydrolysis of gluconolactone to gluconic acid. Significant improvement is expected when gluconolactonase is added to the system. The stability of the sensors differed from 1 to 3 days, with notable sensitivity close to the lower detection limit during the first 24 hours.

The effflcient amplification of the sensor response i.e. the extremely high sensitivity of the enzyme electrode is the basis for the assay of enzymes used as labels in immunoassays as well as the determination of metabolites.

Detection of catecholamines from adrenal chromaffin cells

The secretion of catecholamines in the adrenal tissue is controlled by the splanchnic nerve and influenced by a number of factors, including circulating hormones and paracrine and autocrine mechanisms.

Chromaffin cells secret only epinephrine, norepinephrine and traces of dopamine in the samples. Adrenal chromaffin cells were isolated from adult bovine adrenal medullae and cultivated at a density of 500.000 cells/well. The cells were stimulated, and the medium was collected.

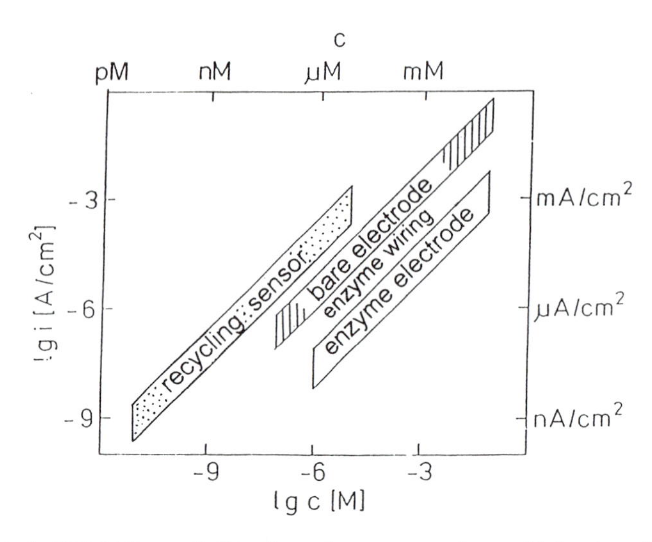

Figure 4. Comparison of the dynamic range and sensitivity for different electrochemical enzyme sensors.

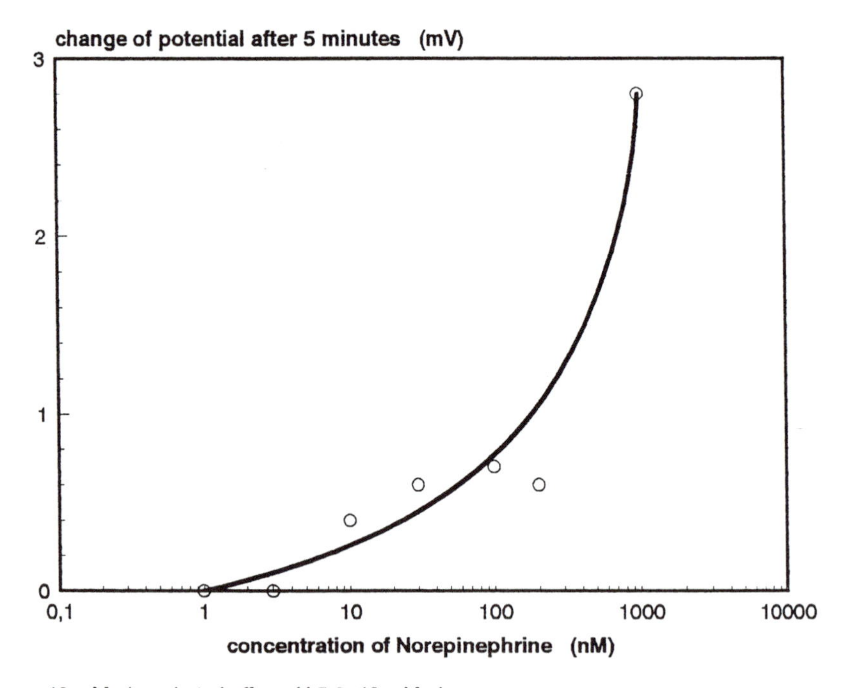

Figure 5. Detection of norepinephrine using the GDH/ Laccase covered pH-ISFET in 10 mM phosphate buffer, pH 5.8 containing 10 mM glucose.

The response of the electrode shows a significant difference between the secretion of stimulated and non-stimulated chromaffin cells. A different catecholamine content of cells after stimulation with nicotinic or nicotine free buffers is also obvious (Figure 6).

It should be noted, that the Lacc-GDH recycling electrode does not distinguish between epinephrine, norepinephrine or dopamine. Nevertheless, the selectivity is sufficient for discriminating the catecholamine signal from other components of the stimulating buffer without any purification or separation step. If the ratio between epinephrine and norepinephrine shall be determined, on the basis of the bioelectrocatalytic substrate regeneration by GDH immobilized on a carbon electrode (15) norepinephrine can be detected 40 times more sensitive than epinephrine while 1 : 4 selectivity is observed for the Lacc-GDH electrode. The conventional HPLC procedure for determination of catecholamines requires complex and expensive devices since the electrochemical detector has only low selectivity.

The analysis based on Lacc-GDH biosensors does not need any pre-treatment of the sample, requires a sample volume of only 10 µl for single analysis and an analysis time of only 5 min.

Indication of Immunoreactions

The cyclic amplification can be connected to the measurement of an immunoreaction. For the determination of antigens the recycling electrode can be used to measure a marker enzyme of a conventional enzyme linked immunosorbent assay (ELISA), which generates a shuttle molecule for the cycle, e.g. aminophenol or ferrocene. The detection of the reaction products formed by alkaline phosphatase is used in different schemes of immunoassay. Heineman and co-workers (16) pioneered the use of PAPP as a substrate of alkaline phosphatase in combination with amperometric measurement of PAP. In this case the detection limit of 7 nM of PAP is approximately 20 times better than the spectrophotometric method. For hapten determination the new dimension of sensitivity of enzyme electrodes offers a novel approach to competitive immunoassays. For this purpose a hapten has to be conjugated with the shuttle molecule. No enzyme label is required. The competition between the analyte hapten and the shuttle hapten for binding to the respective antibody is quantified by the amplification sensor. Since the binding to the antibody makes the shuttle inaccessible to the recycling enzymes, separation and washing steps are no longer necessary.

Sandwich - Immunoassay for IgG

Determination of IgG has been performed in a sandwich type immunoassay using the recycling electrode as detector and both alkaline phosphatase and ß-galactosidase as enzyme label.

As a first model system goat-IgG (gtIgG) was chosen as an analyte The use of alkaline phosphatase as label required a change the pH between the immunoassay (pH 8) and the electrode reaction (pH 6.5). In order to reduce background signal activity determination was performed with 1 µM PAPP. In 0.1 M MES buffer the electrode sensitivity towards PAPP is approximately 10 % of the respective response in the presence of the same concentration of PAP At the same time, the sensitivity towards PAPP in phosphate buffer is 200 times lower in comparison to PAP sensitivity. This suggests that the enzyme membrane causes the hydrolysis of PAPP to PAP. Obviously, phosphate ions inhibit this process.

Since ß-galactosidase (ß-GAL) has the pH optimum in the low acidic range no buffer change was required when this enzyme was used as marker in the immunoassay. The results of the two-step assay (gtIgG + anti-gtIgG-ß-GAL conjugate) are shown in Fig. 7 comparing the photometric (substrate p-nitrophenyl-ß-D-galactopyranoside, PNPGAL) and the bienzyme electrochemical detection schemes, using p-aminophenyl-ß-D-galactopyranoside (PAPGAL) as enzyme substrate. 1 mM PAPGAL was used as substrate resulting in 20 µM in the measuring cell however, much less noise was produced compared to the alkaline phosphatase system and PAP.

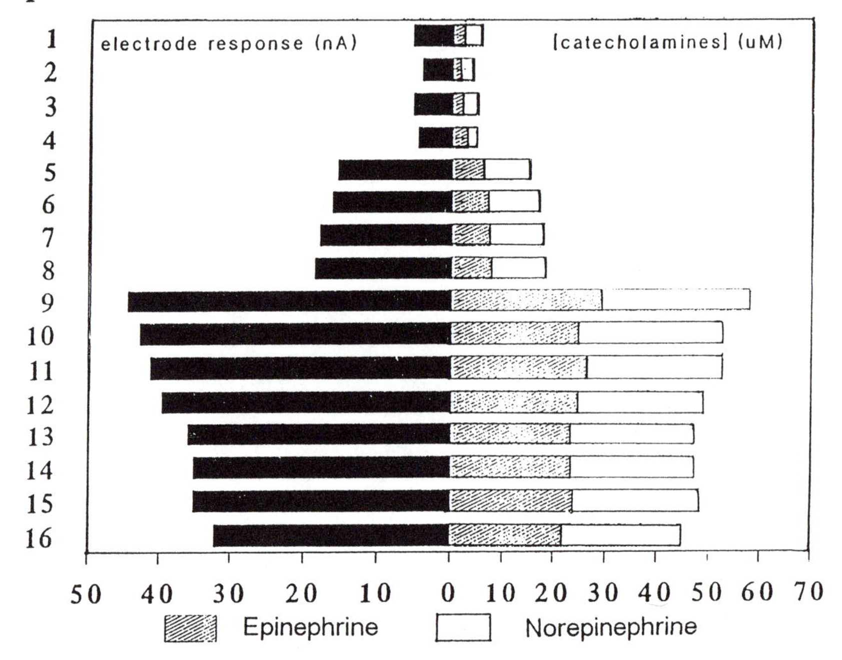

Figure 6. Measurement of catecholamines by the Lacc-GDH electrode. The samples 1 to 4 represent controls with pure incubation buffer (basal response), the samples 5 to 8 represent a stimulation with 30 μM nicotine in incubation buffer (maximal response). The samples 9 to 12 are the lysates of the cells from the samples 1 - 4, the numbers 13 -16 are lysates responding to the samples 5 - 8.

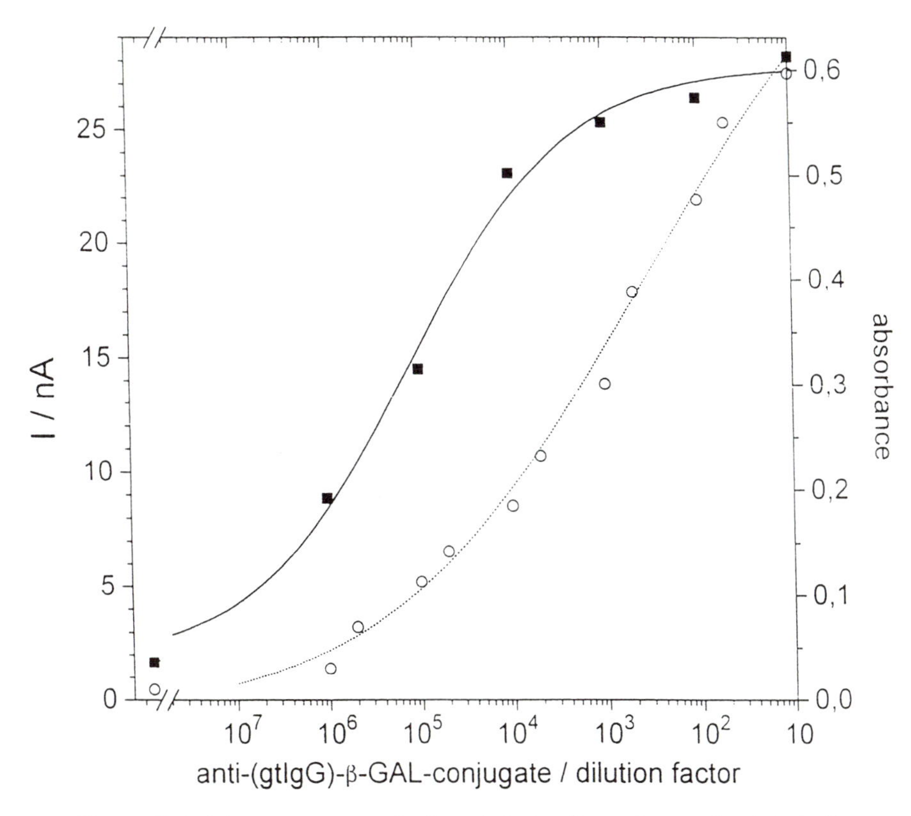

Figure 7. IgG-immunoassay. goat-IgG coated microtiter wells were incubated with various amounts anti-(goat-IgG)-ß-galactosidase conjugate. The enzyme label was determined using either 1 mM PAPGAL or 5 mM PNPGAL, both incubated for 30 min. The response of the bienzyme electrode to 1000 times diluted PAPGAL solution was measured (filled squares, left axis) and compared to the photometric determination using PNPGAL (open circles, right axis).

Detection of Cocaine by Combining Conjugate Displacement and a Substrate-Recycling-Sensor

The cocaine assay is based on a displacement of an enzyme-labeled hapten by the free hapten in the sample. Subsequently the product formed by the enzyme-label triggers the substraterecycling-detector.

Displacement requires two steps: The immobilized antibody is first saturated with labeled antigen. After washing off the excess label the displacement of labeled antigen by antigen (analyte) follows. The displacement assay has been performed in two systems: (i) In ELISA-plates high-affinity monoclonal antibodies against cocaine have been adsorbed and saturated with a cocaine-alkaline phosphatase-conjugate. In the dependence on the cocaine concentration this conjugate is displaced and therefore appearing in the supernatant. The displacement took place during a 15 min. incubation with cocaine. The calibration curve for cocaine (increasing alkaline phosphatase activity in the supernatant) rises between 10 nM and 10 μM cocaine. (ii) In a column-based stopped flow system cocaine concentrations from 1 μM up to 10 mM can be measured. The detection limit is two orders higher than in ELISA-plate assay. That may be due to the dispersion in the flow system and to the shorter incubation time (5 min.). In this system the antibody is immobilized via sugar residues to AvidGel AX. For each injection of analyte the background signal due to non-specific dissociation of label is measured and subtracted. Prior to the first injection of sample the label-saturated column remaining free antibody binding sites have been saturated with cocaine. The alkaline phosphatase concentrations of all the samples have been measured off-line using the amplification enzyme sensor.

Redox label Immunoassay

The ability to determine nanomolar concentrations of ferrocene derivatives allows for the use of ferrocene-conjugated haptens in an immunoassay. In this case, antibodies present in the reaction media lead to a decrease in the electrode response. This is mainly due to the antibody-ferrocene complex not being able to diffuse through the dialysis membrane into the Lacc-GDH membrane. A model immunoassay system was developed using antibodies against ferrocene-benzoic acid isothiocyanate conjugate (fer-benz). Fer-benz is a good substrate for the Lacc-GDH sensor and its detection limit was 0.5 nM. After incubation of 500 nM fer-benz with different amount of antibodies for 15 min., the solution was transferred to the measuring cell (100-fold dilution, final concentration of 5 nM). The increase in the concentration of antibodies resulted in a decrease in the electrode response.

The effect of the antibody on the electrode response was also examined without preincubation. In this case antibodies were injected into the cell and the response on 10 nM substrate was indicated. The presence of antibodies against fer-benz in the cell leads to a decrease in the electrode response. At a low antibody dilution there is no response. Additionally, antibodies raised against fer-benz had no influence on the electrode response to p-aminophenol and weakly influences the response to ferrocendicarboxylic acid. Unspecific antibodies (against FITC) also had negligible influence on the response to fer-benz. This confirms that the specific antigen-antibody interaction causes the changes in electrode response.

The results indicate that the application of the ferrocene derivatives for labeling haptens in combination with the recycling Lacc-GDH electrode provides a promising way forward for the development of immunoassays within the nanomolar concentration range. The main advantage of the present approach is the very short (several seconds) assay time.

Acknowledgments

The authors acknowledge the financial support from the BMFT-projects: EUREKA (FKZ: 0319579A),and Adrenalin (FKZ: 0310821 and FKZ: 0310822).

References

1. E.H.Fischer, J.U.Becker, H.E.Blum, C.Heizmann, G.W.Kerrick, P.Lehky, D.A.Malencik & S.Poinwong in: Molecular Basis of Motility. Springer Verlag Berlin-Heidelberg, 137-158 (1976).
2. 0.H.Lowry & L.V.Passonneau in: A Flexible System of Enzymatic Analysis. Chapter 8. Academic Press, New York & London, 129-145 (1972).
3. F.Schubert, D.Kirstein, K.-L.Schroder & F.W.Scheller. Anal. Chim. Acta, 169, 391-396 (1985).
4. E.H.Hansen, A.Arndal & L.Norgaard. Anal. Lett., 23, 225-240 (1990).
5. F.Schubert, D.Kirstein, F.Scheller, R.Appelqvist, L.Gorton & G.Johansson. Anal. Lett., 19,1273-1288 (1986).
6. T.Yao, N.Kobayashi, T.Wasa. Electroanal., 2, 563-566 (1990).
7. F.Schubert & F.Scheller. Methods Enzymol., 137, 152-160 (1988).
8. F.Mizutani, Y.Shimura & K.Tsuda. Chem. Lett., 199,199-202 (1984).
9. F.Scheller, N.Siegbahn, D.Danielsson & K.Mosbach. Anal. Chem., 57,1740-1743 (1985).
10. F.Mizutani, T.Yamanaka, Y.Tanabe & K.Tsuda. Anal. Chim. Acta, 177,153 (1985).
11. U.Wollenberger, F.Schubert, D.Danielsson & K.Mosbach. Stud. Biophys., 119, 167-170 (1987).
12. C.F.Thurston. Microbiology, 140,1 9-26 (1 994).
13. A.Sucheta, B.A.C.Ackrell, B.Cochran & F.A.Armstrong. Nature 356, 361-362 (1992).
14. A.I.Yaropolov, V.Malovik, S.D.Varfolomeyev & I.V.Berezin. Dokl. Akad. Nauk SSSR, 249, 1399-1401 (1979).
15. A.Eremenko, A.Makower, W.Jin, P.Rieger & F.W.Scheller. Biosensors & Bioelectronics, in press.
16. H.T.Tang, C.F.Lunte, H.B.Halsall &W.R.Heinemann. Anal. Chim. Acta, 214,187-195 (1988).

RECEIVED September 15, 1995

Chapter 8

Electroenzymatic Sensing of Fructose Using Fructose Dehydrogenase Immobilized in a Self-Assembled Monolayer on Gold

K. T. Kinnear and H. G. Monbouquette

Chemical Engineering Department, University of California, Los Angeles, CA 90095–1592

The hydrophobic enzyme, fructose dehydrogenase (from *Gluconobacter* sp., EC 1.1.99.11), and coenzyme Q_6 have been co-immobilized in a self-assembled monolayer (SAM) on gold through a detergent dialysis procedure to create a prototype fructose biosensor. The SAM consists of a mixture of octadecyl mercaptan (OM) and two short chain disulfides, which form -S-CH_2-CH_2-CH_2-COO^- and -S-CH_2-CH_2-NH_3^+ on the surface. The short chain, charged modifiers may provide defects, or pockets, in the OM layer where the enzyme may adsorb through electrostatic interactions. At oxidizing potentials, the electrode generates a catalytic current at densities up to about 10 $\mu A/cm^2$ when exposed to fructose solution. The enzyme electrode exhibits a response time well under a minute and the calibration curve is linear at fructose concentrations up to 0.8 mM. The biosensor prototype exhibits low susceptibility to positive interference by ascorbic acid indicating that this construct could be useful for fructose analysis of citrus fruit juice.

A reliable fructose sensor would be of value for the quantitation of this sugar in fruit juice, wine, corn syrup, blood serum, and seminal plasma. Note that a low level of fructose in seminal plasma can be an indicator of male infertility. Several reports have appeared describing the development of prototype amperometric fructose biosensors using *Gluconobacter* sp. fructose dehydrogenase (EC 1.1.99.11) (*1-5*), a 140 kDa pyrroloquinoline quinone (PQQ)-containing oxidoreductase. The enzyme has been immobilized on gold, platinum and glassy carbon (*3*), entrapped in conductive polypyrrole matrices (*4,5*), or confined to the surface of a carbon paste electrode behind a dialysis membrane (*1,2*). Although sensors exhibiting good current density, sensitivity, stability and response time have been described, erroneous signal generation due to electroactive interferents such as ascorbic acid in citrus juice has been a persistent problem. This paper describes an effective approach to the problem of polar, ascorbic acid interference by embedding the enzyme in a mostly hydrophobic, insulating self-assembled monolayer on gold.

The hydrophobic nature of membrane-bound redox enzymes, such as this fructose dehydrogenase, provides an obvious route to gentle, stabilizing immobilization in hydrophobic adsorbed layers on electrodes which mimic the cell

0097–6156/95/0613–0082$12.00/0

membrane microenvironment. This concept has been demonstrated earlier with the stable immobilization of membrane-bound *E. coli* fumarate reductase in an alkanethiolate layer on gold (*6*). In this case, direct electron transfer was achieved between the enzyme and electrode. In contrast, stable immobilization of fructose dehydrogenase required the co-adsorption of short-chain, charged disulfides for electrostatic binding of the enzyme (*7*); and a mediator, coenzyme Q_6, was needed for electron transfer. Although others have achieved direct electron transfer between unmodified electrode surfaces and this fructose dehydrogenase (*2,3*), some of the preparations are very unstable (*3*). We report a stable and highly electroactive membrane mimetic system where coenzyme Q_6 mediates electron flow between the enzyme and electrode.

Experimental

Reagents and Materials. D-fructose dehydrogenase (FDH) from *Gluconobacter* sp., coenzyme Q_6, D(-)fructose, and *n*-octyl β-D-glucopyranoside (*n*-octyl glucoside) were purchased from Sigma and were used without further purification. Octadecylmercaptan, cystamine dihydrochloride, and 3,3'-dithiodipropionic acid were obtained from Aldrich. Other chemicals were reagent grade from Fisher or Sigma and were used as received. All electrochemical supplies were purchased from Bioanalytical Systems, Inc. (W. Lafayette, IN).

Electrochemical Procedure. Electrochemical data was obtained with a BAS CV-1B instrument and a BAS X-Y recorder. The BAS CV-1B was interfaced to a Macintosh IIcx with a National Instruments (Austin, TX) Lab-NB board and software (LabVIEW II). Gold disk electrodes (d = 1.6 mm, $A_{geom} = 0.02$ cm^2), a Ag/AgCl (3 M NaCl) electrode and a platinum wire were used as the working, reference and counter electrode, respectively. Before each experiment, a gold electrode was polished and electrochemically cleaned as described earlier (*6,8*). The amperometric determination of fructose was done at 0.5 V vs. Ag/AgCl in stirred deoxygenated 10 mM KH_2PO_4, pH 4.5 under a blanket of Ar at room temperature unless otherwise stated.

Adsorbed Layer Preparation. A stock solution of FDH (~1.35 μM) was prepared with 2.5 mg of FDH and 1 ml of phosphate buffer and 1% octyl glucoside (35 mM) and used within two weeks (stored at 4 °C between experiments). Coenzyme Q_6 at 170 μM was added to the enzyme detergent solution when indicated. Prior to dialysis, freshly polished and electrochemically cleaned electrodes were modified for 2 hours in various thiol/disulfide ethanol solutions. In addition to octadecyl mercaptan (OM), two disulfides were used, cystamine dihydrochloride (CA) and 3,3'-dithiodipropionic acid (TP). The disulfides undergo dissociative adsorption at the gold surface and form self-assembled monolayers in much the same way as alkanethiols (*2*). The total thiol/disulfide concentration was 1 mM with 40% OM, 30% TP, and 30% CA. The FDH stock solution (270-280 μl) and the modified electrode were added to a 10,000 MWCO dialysis bag and placed in a 400 ml reservoir of phosphate buffer; dialysis was carried out at 4 °C with stirring. The dialysate was replaced 3 to 4 times over 18 to 48 hours; the last reservoir replacement included approximately 1 ml of the detergent-sorbing resin, Calbiosorb. The electrode was removed from the bag and rinsed with pure water and either used immediately or stored in buffer.

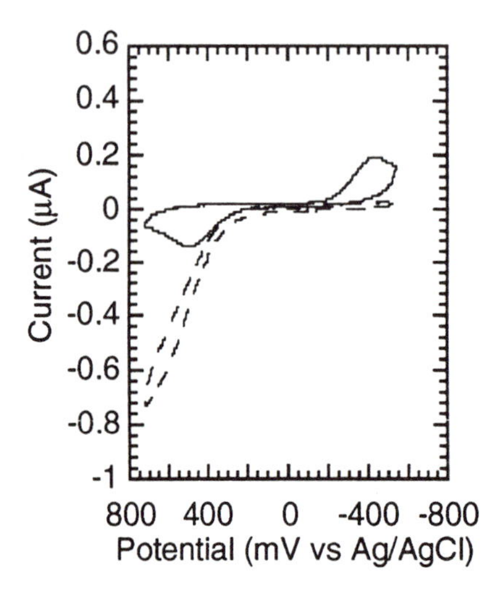

Figure 1. Steady state CV of FDH and Q_6 in a mixed thiol monolayer on gold in deoxygenated 10 mM KH_2PO_4, pH 4.5 buffer (solid curve); and the first CV scan in 28 mM fructose (dashed curve). The scan rate was 50 mV/s.

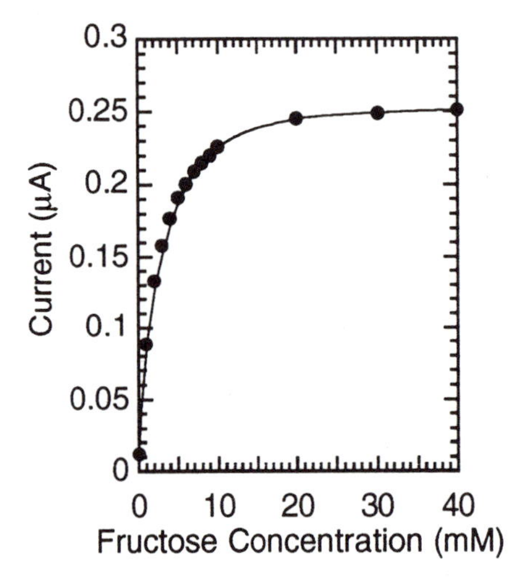

Figure 2. Calibration curve for fructose obtained with an FDH-Q_6 modified electrode.

Results and Discussion

Coenzyme Q_6 (Q_6) and fructose dehydrogenase were immobilized simultaneously through detergent dialysis on a gold electrode previously modified using a 40% OM, 30% TP and 30% CA mixture. A strong catalytic current response was observed when this electrode was introduced into fructose solution (Figure 1). Control voltammograms (CVs) of a Q_6 mixed monolayer electrode without enzyme do not show catalytic activity in the presence of fructose (data not shown). The enzyme could not be immobilized in an OM layer without TP and CA. As has been demonstrated earlier (*7*), these charged species can enable stable electrostatic binding of redox proteins to gold.

To investigate the potential of this system for sensing applications, various parameters were obtained. Figure 2 depicts calibration curves obtained with a Q_6-FDH modified electrode. At lower concentrations of fructose, a linear relationship between current and fructose concentration is observed (Figure 2). Saturation values of current are observed at higher fructose concentrations. The current density of this electrode at 10 mM fructose is 11.3 $\mu A/cm^2$ versus 6.9 $\mu A/cm^2$ for an FDH carbon paste dialysis membrane electrode reported by Ikeda *et al.* (*2*).

The time response of the Q_6-FDH modified electrode was also evaluated as shown in Figure 3. Upon addition of 5 mM fructose to a stirred phosphate buffer solution, the maximum steady-state current response was obtained within about 20 seconds.

Another important biosensor feature to evaluate is susceptibility to electroactive interferents. In FDH electrode applications, such as the measurement of fructose in fruit juices or wine, the effects of ascorbic acid should be investigated. Ascorbic acid is oxidized quite readily at bare electrode surfaces. The blocking mercaptan layer of this system may have the advantage of hindering access of ascorbic acid to the electrode surface. Figure 4 illustrates the effect of 100 μM ascorbic acid on the measured current of a 2 mM fructose solution. The presence of ascorbic acid at 5% of

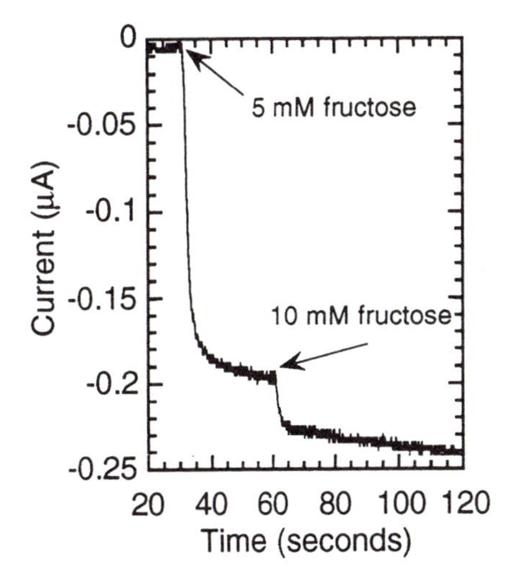

Figure 4. Effect of ascorbic acid on current response obtained with an FDH-Q_6 modified electrode.

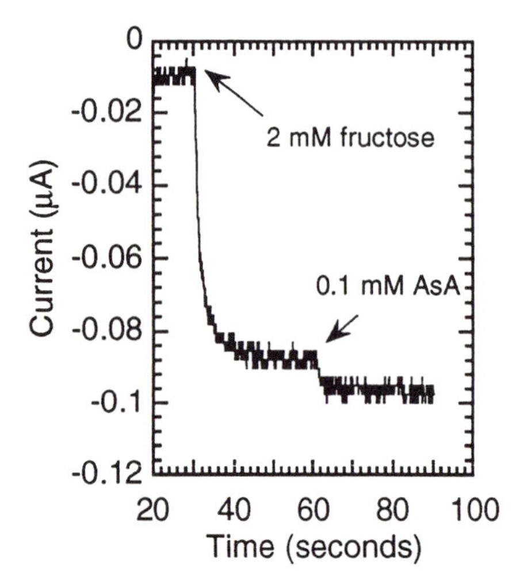

Figure 3. Time dependence of the current response obtained with an FDH-Q_6 modified electrode.

the fructose concentration results in an error of 9%. Note that ascorbic acid usually is 2-3% of the fructose concentration in citrus fruit juices. The presence of the mixed monolayer does seem to block access of ascorbic acid to some degree. In contrast, the FDH carbon paste electrode of Ikeda *et al.* gives an error of approximately 80% at 50 μM ascorbic acid and 2 mM fructose (*2*).

The membrane mimetic fructose dehydrogenase system for fructose sensing described in this paper is stable for at least several days. Activity loss appears to occur primarily due to desorption of Q_6 from the electrode surface since much electrode activity could be restored by exposure of the electrode to a solution containing an alternate lipophilic mediator, decylubiquinone. Efforts are underway to improve mediator retention with the hope that electrode stability can be increased further to a practical level.

Acknowledgments

This research was supported by National Science Foundation grant BES-9400523.

Literature Cited

1. Ikeda, T.; Matsushita, F. and Senda, M. *Agric. Biol. Chem.* **1990,** *54,* 2919.
2. Ikeda, T.; Matsushita, F. and Senda, M. *Biosens. Bioelectron.* **1991,** *6,* 299.
3. Khan, G.F.; Shinohara, H.; Ikariyama, Y. and Aizawa, M. *J. Electroanal. Chem.* **1991,** *315,* 263.
4. Khan, G.F.; Kobatake, E.; Shinohara, H.; Ikariyama, Y. and Aizawa, M. *Anal. Chem.* **1992,** *64,* 1254.
5. Begum, A.; Kobatake, E.; Suzawa, T.; Ikariyama, Y. and Aizawa, M. *Anal. Chim. Acta* **1993,** *280,* 31.
6. Kinnear, K.T. and Monbouquette, H.G. *Langmuir* **1993,** *9,* 2255.
7. Hill, H.A.O. and Lawrance, G.A. *J. Electroanal. Chem.* **1989,** *270,* 309.
8. Finklea, H.O.; Avery, S.; Lynch, M. and Furtsch, T. *Langmuir* **1987,** *3,* 409.

RECEIVED June 5, 1995

Bioprocess Monitoring and Control

Chapter 9

Recent Advances in Bioprocess Monitoring and Control

Weichang Zhou[1] and Ashok Mulchandani[2]

[1]Bioprocess Research and Development Department, Merck Research Laboratories, Merck & Co., Inc., P.O. Box 2000, Rahway, NJ 07065
[2]Chemical Engineering Department, College of Engineering, University of California, Riverside, CA 92521

Recent advances in analytical instrumentation, biosensors and computer technology have lead to the development of new approaches and techniques to control and optimize bioprocesses. Use of on-line measurements to monitor bioprocesses for identifying physiological state of cells and controlling environmental conditions to maintain the cells in the optimal physiological state has become more prevalent. Novel control strategies, along with new concepts, such as expert systems and artificial neural networks have been introduced and applied in bioprocess monitoring and control. The bottleneck, however, is the unavailability of simple, reliable and robust on-line and *in situ* measurement probes and on-line sampling devices, especially for industrial applications. Further development in this field will strongly depend on the improvement of these basic on-line tools.

The rapid development of recombinant microbial and cell culture technology in the last decade has lead to rapid growth and numerous innovations in the field of bioprocess monitoring and control. Process optimization to reduce the manufacturing cost of increased number of biologically manufactured products continues to drive the development of novel on-line measurement techniques and implementation of practical control strategies to achieve high performance cultures. Ideally, a control strategy would allow one to maintain the cultured cells at their optimal physiological state so that both high productivity and product quality are ensured. In order to maintain optimal physiological state, the medium composition and environmental conditions at different culture stages have to be manipulated differently. This will require dynamic control of the nutrient feeding to the bioreactor. For example, to obtain high cell

0097–6156/95/0613–0088$12.00/0

concentration in fed-batch mammalian cell cultures, dynamic control of nutrient feeding is necessary to maintain glucose and essential amino acids at low balanced levels to reduce the production of inhibitory metabolites, such as lactate and ammonium.

Besides the traditional research activities of developing novel biosensors and methods, the application of existing techniques to monitor and control bioprocesses has attracted increasing attention, especially in cell culture processing [1]. The application of expert systems and hybrid models, which use all the available knowledge about a specific process, to supervise process control in a situation where mathematical models alone are not able to describe the entire process, is another area rapidly gaining interest. Expert systems have found wide applications in monitoring and controlling microbial fermentation [2] and have been recently proposed for monitoring and controlling more complicated cell culture processes [3]. The expert system approach has been also used for supervising on-line measurements, such as a flow injection analysis (FIA) system [4].

Because there are many comprehensive reviews available on measurements, instrumentation and control strategies in this field, especially in the microbial fermentation area [2, 5-10], this article will be devoted to providing an overview on the recent developments with the emphasis in on-line measurements and control concepts for monitoring and controlling cell culture processes. In comparison with microbial process monitoring and control, which experienced rapid development in the 1980's [7], monitoring and control of animal cell culture processes has been witnessing a significant advancement since just the early 1990's.

There are two essential components in any bioprocess monitoring and control scheme: (1) suitable *in situ* or *ex situ* sensors for on-line measurement of desired parameters and (2) a control system with the necessary software and hardware for a suitable control strategy. Specific examples of recent developments in on-line measurements and control concepts will be highlighted in the following sections to reflect the rapid progress in this field.

On-line Measurements

On-line measurements are the key to implementing any kind of control strategies. Despite numerous efforts towards development of *in situ* biosensors, their use in bioreactors is still limited. Currently, only a few parameters, such as temperature, pH and dissolved oxygen concentration are measured using *in situ* probes. However, significant advances in developing and applying the new on-line measurement techniques have been achieved recently.

Optical density probes

Cell concentration, the most important parameter in most biological processes, can be determined on-line using various techniques. Even though there are numerous on-line and off-line measurement techniques, the common reliable method is still off-line determination of dry cell mass weight, where possible [11]. *In situ* optical density (OD) probes installed in bioreactors are the simplest of the various on-line measurement techniques available. Recently, attempts have been made to measure cell concentration on-line using different kinds of OD probes [12-17]. These probes work on the principle of light transmission, or back scattering and use different light sources including laser lights with a precise wavelength to detect culture turbidity, which is correlated with cell concentration. Very successful results using these probes have been reported for on-line determination of total cell concentration in microbial [12-13], animal [14-16] and plant cell cultures [17]. Even though this measurement does not provide information about cell viability, it does provide some very useful insight about the physiological state of cells in combination with other on-line measurements [15]. For example, OD probes have been used to determine the specific growth rate, and in combination with oxygen uptake rate measurement to determine specific metabolic parameters. OD probes have also been used to control nutrient feeding rates in fed-batch and continuous perfusion cultures [12, 13, 16]. However, it is not possible to obtain microscopic information, such as cell size and cell shapes using these OD probes. Since they are simple and non-invasive, it is expected that these probes will become a standard tool in both industrial manufacturing processes and academic research.

Image analysis

To overcome the drawbacks of the OD probes, image analysis systems coupled with *in situ* microscopy have rapidly emerged as a technique for determining both viable and non-viable cell concentration and for characterizing cell populations in bioreactors [18-20]. Recent advancements in computer technology has made image processing relatively fast to allow on-line determination of microscopic information such as cell size and morphology. In this technique, ideally, a microscope is directly mounted in a port of a reactor through the tube of a front-end sensor housing and *in situ* images of the cells are obtained for visual and automatic examination of morphological and statistical parameters. A fluorescence microscope can be used for differentiation between viable and non-viable cells. This approach was successfully used to determine viable and non-viable cell concentration in a yeast fermentation [19]. Differentiation of viable and non-viable cells can also be achieved by measuring the difference of brightness distribution of viable and trypan blue stained non-viable cells. For this purpose, samples are pumped into a chamber, where they are stained with trypan blue and images are taken. For non-destructive measurements, the particle size can be used to distinguish viable and non-viable cells, since viable cells are typically larger than non-viable cells in anchorage-independent mammalian cell cultures. The microscopic image is then processed using a computer system [18]. A similar image analysis system has also been successfully used to differentiate and classify somatic embryos into different morphological groups in a plant cell culture using a pattern recognition system. This system allows one to follow kinetics of embryo development on both population and individual levels and to evaluate the quality of embryos [20].

Flow injection analysis and HPLC

On-line measurements of medium composition and metabolite concentrations allow one to control critical nutrients at desired levels. However, the realization of the dream of their on-line determination as well as on-line measurements of product concentration and cell concentration in anchorage-dependent mammalian cell cultures by simply installing a bunch of biosensors in bioreactors still seems far away. Despite numerous efforts to develop *in situ* biosensors, their use in bioreactors is still limited. Recent efforts have been directed towards using on-line measurement of these parameters outside the bioreactor by (a) sensors coupled to flow injection analysis (FIA) and (b) high pressure liquid chromatography (HPLC) techniques [9]. These measurements, which are performed *ex situ*, require continuous sample that can be provided by on-line sampling devices. These devices, which are based on membrane filtration [9, 21, 22], can be installed *in situ* or *ex situ* to provide 1-2 mL/min of sample for external analyzers. *In situ* sampling devices are preferred over *ex situ*, because of the following reasons: (1) since an *ex situ* device requires a pump to maintain a high circulation rate across the filter membrane, the cells are subjected to mechanical shear which may affect their viability, (2) during the period the cells are outside the bioreactor, they may suffer from oxygen or nutrient limitation which may alter their metabolic activity and (3) risk of contamination. Regardless, there are also operational problems when using either *in situ* or *ex situ* membrane devices. With continued usage, these membrane devices also begin to reject analytes of interest due to changing pore size. For example, as the membrane fouls over a long period of use, larger molecular weight compounds such as therapeutic proteins cannot pass through completely. Further development of a simple and reliable sampling device will accelerate the implementation of external on-line analysis.

FIA is a very simple and elegant technique that is rapidly finding applications in on-line bioprocess monitoring [9, 22-24]. In FIA, a sample to be analyzed is injected into a moving non-segmented carrier stream of buffer or reagent and reacts with the carrier stream in the manifold, as it is transported toward a detector which continuously records the change in either absorbance, pH, dissolved oxygen, conductivity or any other physical parameters. The typical output of a FIA is a peak because of the dispersion of the injected sample zone. The concentration is related to the peak height or the area under the peak. Essential components of a FIA system are: (a) a selector valve used to switch flow between a sample to be analyzed, calibration standards or wash/purge solutions; (b) a multi-channel peristaltic pump module; (c) an injection valve used to load a specific volume of sample or standard (20 to 50 μL) into a carrier stream at programmed intervals, (d) a manifold, where reactions take place and (e) a detector/transduction system with a recorder or computer. Since the residence time of a sample in FIA system is typically less than 30 seconds, at least 120 samples per hour can be analyzed. FIA systems have been applied for on-line analysis of numerous low molecular weight compounds in the culture broth, such as glucose, lactate, glutamine and phosphate [9, 25-34] and even proteins using turbidimetric immunoassay or heterogeneous assay with a working principle similar to affinity chromatography [22-24]. Recent developments in FIA systems include further refinement by including on-line calibration based on internal standard principle [35] and improving the robustness, reliability and automation [4, 24]. An expert system was introduced to automate and supervise the FIA operations [4].

On-line HPLC has been shown to be useful for determination of fermentation nutrient and product concentrations for feed-back control [36, 37]. In comparison with FIA, HPLC gives more accurate and reliable measurements, however, the frequency of analyses is generally low. Additionally, an HPLC system is more expensive than a FIA system.

On-line off-gas analysis for determination of OUR, CPR and RQ

Oxygen uptake rate (OUR), carbon dioxide production rate (CPR) and respiration quotient (RQ, RQ=OUR/CPR) are very important metabolic parameters in biological processes that can be used to obtain information about the physiological state of the culture. On-line determination of these parameters by gas and liquid phase balances through the analysis of exhaust gas composition using infrared CO_2 and paramagnetic O_2 analyzer or mass spectrophotometer is very widespread in microbial fermentation processes [38]. However, the application of on-line exhaust gas analysis is rather limited in mammalian cell cultures, since it is insensitive and inefficient (1) due to small oxygen consumption and high gas flow rate required to maintain the desired dissolved oxygen (DO) level, especially in surface aerated bioreactors, (2) large gas hold up in the head space and (3) use of bicarbonate-buffered medium. The problems of high aeration rate and gas hold up have been recently resolved by developing optimized aeration methods where surface aeration was combined with direct sparging. For example, gas with a high O_2 content is used for sparging while air is used for surface aeration. The problem of CO_2 released from bicarbonate in the medium in the calculation of CPR has been solved through the use of models to calculate the portion of CO_2 released from bicarbonate. Recently, on-line gas analysis has been successfully performed using both infrared CO_2 and paramagnetic O_2 [39, 40] analyzers or a mass spectrometer [41-44].

As an alternative to on-line gas analysis, simple and reliable techniques such as dynamic measurements [15, 45], stationary measurements [46] and the recirculation method [47] have been developed for on-line OUR determination in mammalian cell cultures. In the dynamic method, a DO profile is dynamically created by changing the gas composition. The DO concentration is increased to a first DO set point by injecting pure O_2 or air. Subsequently, nitrogen gas is used to flush the O_2 from the bioreactor head space to lower the O_2 concentration to a second DO set point. The time profile of DO between the two set points is used to calculate the OUR. This method is very sensitive, even though it does not provide continuous OUR data. However, the DO varies in a big window. Such a variation of DO may affect cell growth and product formation, especially when the normal DO setpoint is very high or very low.

The stationary method allows the determination of OUR continuously. In this method, the DO concentration is controlled at a constant level by varying the O_2 composition in the gas phase by varying the flow rate of the gases, O_2, CO_2 and N_2. OUR is calculated through an O_2 balance in the liquid phase. Although this method provides continuous OUR data, it is not very sensitive, especially in batch cultures, when OUR is relatively small. In both methods, the K_La value for oxygen transfer is either predetermined and assumed to be constant or can be adjusted on-line.

In the recirculation method, the culture broth is circulated through a recycle loop, located outside the bioreactor, at a certain flow rate. The difference between the DO concentrations in the bioreactor and the external loop is proportional to the pumping rate and the OUR of the culture. This method requires (1) optimization of the pumping rate to provide a relatively large difference between the two DO concentrations without causing an O_2 limitation for the cells and (2) accurate determination of the external loop volume.

Control Concepts

Advanced control concepts, such as adaptive control strategies using mathematical models, expert systems and artificial neural networks have been widely applied for real-time fault diagnosis, estimation of unmeasurable parameters, on-line prediction and state estimation in microbial fermentation processes [2, 7]. In contrast, such control concepts are waiting to be introduced in animal cell culture processes [1]. Even though the expert system approach has been proposed to monitor and control animal cell culture processes, because of the complexity of the culture medium and cellular metabolism, currently these processes are generally monitored manually. Nevertheless, various control strategies for maintaining nutrients within critical ranges by dynamically adjusting the feeding rates have been developed and applied to obtain high performance cultures. Kurokawa et. al. [37] controlled glucose and glutamine concentrations at low levels in order to reduce the production of inhibitory metabolites, such as lactate and ammonium, in fed-batch hybridoma cultures by controlled feeding of concentrated glucose and glutamine solutions. A much higher cell concentration was obtained by this method when compared with a batch culture. For the feed-back control, the glucose and glutamine concentrations were measured on-line using an HPLC. OUR was used to control feeding of a nutrient concentrate in fed-batch hybridoma cultures based on stoichiometric relations among consumption of oxygen and other critical nutrients [48]. The feeding rate of a salt-free nutrient concentrate was calculated on-line based on OUR measurements. By controlling glucose concentration at a low level, amino acid concentrations were also maintained at low levels. The cell metabolism was altered, resulting in a more oxidative cellular metabolism and reduction of metabolite production. Cells grew to a very high concentration with a high viability. At the end of the growth stage, a maximal cell concentration of $1.3x10^7$ cells /mL with a viability of > 95 % was obtained in compared with a cell concentration of less than $2x10^6$ cells/mL in a batch culture. Such a control strategy was also used to adjust perfusion rate in a recombinant human kidney epithelial 293 cell culture [49]. The perfusion rate was adjusted based on on-line OUR measurements to provide cells with sufficient nutrients and to remove metabolites from the reactor. As a result of this dynamic adjustment of the perfusion rate, a very high cell concentration of about $1x10^8$ cells/mL and high productivity were achieved.

Since biological processes are generally very complicated, they have not been fully described using mathematical models. However, hybrid process modeling based on all the available knowledge and information provided by means of mathematical models, heuristic knowledge or even by data sets has been shown to be useful for process state estimation, prediction and control [50, 51].

Conclusions

With the availability of new on-line measurement tools and control concepts, the field of bioprocess monitoring and control has advanced to a new stage. With the existing on-line measurement tools and control concepts, one can develop a reliable control strategy, which will allow one to maintain cells in culture at a physiological state with high performance, by utilizing all available process knowledge. However, on-line tools need further improvement and refinement. Future challenges are primarily associated with making on-line measurements simple, reliable and robust. For *ex situ* sensors, on-line aseptic sampling devices are crucial. *In situ* sample systems will be especially important for industrial applications to eliminate contamination risks. Further development of bioprocess monitoring and control will strongly depend upon the improvement of these basic on-line tools.

References

1. Hu, W.-S. and Piret, J.M. (1992) Mammalian cell culture process. Curr. Opinion Biotechnol., 3: 110-114
2. Konstantinov, KB and Yoshida T. (1992) Mini review: Knowledge-based control of fermentation processes. Biotechnol. Bioeng., 38: 665-677

3. Konstantinov, K. B., Zhou, W., Golini, F., and Hu, W.-S. (1994) Expert systems in the control of animal cell culture processes: Potentials, functions, and perspectives. Cytotechnol., 14: 233-246
4. Hitzmann, B., Gomersall, R., Brandt, J., van Putten, A. (1995) An expert system for the supervision of a multi channel flow injection analysis system. In: Recent Advances in Biosensors, Bioprocess Monitoring, and Bioprocess Control, K.R. Rogers, A. Mulchandani, W. Zhou, Eds., ACS Symposium Series, American Chemical Society, Washington, D.C. (this vol.)
5. Glacken, M.W. (1991) Bioreactor control and optimization. In: Animal Cell Bioreactors, C.S. Ho, D.I.C. Wang, Eds., Butterworth-Heinemann, Stoneham, MA
6. Loch, G., Sonnleitner, B., and Fiechter, A. (1992) On-line measurements in biotechnology: Techniques. J. Biotechnol., 25: 23-53
7. Schügerl, K. (1991) Measuring, Modelling and Control, Biotechnology, Vol. 4, VCH, New York
8. Schügerl, K. (1991) Common instruments for process analysis and control. pp. 5-25. In: Biotechnology, Vol. 4, K. Schügerl, Ed., VCH, New York
9. Schügerl, K. (1991) On-line analysis of broth. pp. 149-180. In: Biotechnology, Vol. 4, K. Schügerl, Ed., VCH, New York
10. Scheirer, W. and Merten, O.-W. (1991) Instrumentation of animal cell culture reactors. pp. 405-443. In: Animal Cell Bioreactors, C.S. Ho, D.I.C. Wang, Eds., Butterworth-Heinemann, Stoneham, MA
11. Reardon, K.F., Scheper, T.H. (1991) Determination of cell concentration and characterization of cells. pp. 179-223. In: Biotechnology, Vol. 4, K. Schügerl, Ed., VCH, New York
12. Yamane, T., Hibino, W., Ishihara, K., Kadotani, Y., Kominami, M. (1992) Fed-batch culture automated by use of continuously measured cell concentration and cell volume. Biotechnol. Bioeng., 39: 550-555
13. Yamane, T. (1993) Application of an on-line turbidimeter for the automation of fed-batch cultures. Biotechnol. Prog. 9: 81-85
14. Konstantinov., K. B., Pambayun, R., Matanguihan, R., Yoshida, T., Perusich, C. M., Hu, W.-S. (1992) On-line monitoring of hybridoma cell growth using a laser turbidity sensor. Biotechnol. Bioeng., 40: 1337-1342
15. Zhou, W. and Hu, W.-S. (1994) On-line characterization of a hybridoma cell culture process. Biotechnol. & Bioeng., 44: 170-177
16. Wu, P., Ozturk, S. S., Blackie, J. D., Thrift, J. C., Figueroa, C., and Daveh, D. (1995) Evaluation and applications of optical cell density probes in mammalian cell bioreactors. Biotechnol. Bioeng., 45: 495-502
17. Zhong, J-J., Fujiyama, K., Seki, T., and Yoshida T. (1993) On-line monitoring of cell concentration of perilla frutescens in a bioreactor. Biotechnol. Bioeng., 42: 542-546
18. Maruhashi, F., Murakami, S., and Baba, K. (1994) Automated monitoring of cell concentration and viability using an image analysis system. Cytotechnol. 15: 281-289

19. Suhr, H., Wehnert, G., Schneider, K., Geissler, P., Jähne, B., Scheper, T. (1995) *In situ* microscopy for on-line characterization of cell-populations in bioreactors, including cell-concentration measurements by depth from focus. Biotechnol. Bioeng., 47: 106-116
20. Chi, C-M., Vits, H., Staba, E.J., Cooke, T.J., and Hu, W-S. (1994) Morphological kinetics and distribution in somatic embryo cultures. Biotechnol. Bioeng., 44: 368-378
21. Schügerl, K. (1990) Aseptic sampling. pp. 1188-1193. In: Proc. 5th Eur. Congr. Biotechnol., C. Christiansen, L. Munck and J. Villadsen, Eds., Vol. II, Munksgaard Int. Publisher, Copenhagen
22. Feng, C., Fraune, E., Freitag, R., Scheper, T. and Schügerl, K. (1991) On-line monitoring of monoclonal antibody formation in high density perfusion culture using FIA. Cytotechnol., 6: 55-63
23. Schulze, B., Middendorf, C., Reinecke, M., and Scheper, T. (1994) Automated immunoanalysis systems for monitoring mammalian cell cultivation processes. Cytotechnol. 15: 259-269
24. Hitzmann, B., Schulze, B., Reinecke, M., Scheper, T. (1995) The automation of two immun-FIA-systems using the flexible software system CAFCA. In Recent Advances in Biosensors, Bioprocess Monitoring, and Bioprocess Control, K.R. Rogers, A. Mulchandani, W. Zhou, Eds., ACS Symposium Series, American Chemical Society, Washington, D.C. (this vol.)
25. Garn, M., Gisin, M., and Thommen, C. (1989) A flow injection analysis system for fermentation monitoring and control. Biotechnol. Bioeng., 34: 423-428
26. Renneberg, R., Trott-Kriegeskorte, G., Lietz, M., Jaeger, V., Pawlowa, M., Kaiser, G., Wollenberger, U., Schubert, F., Wagner, R., et al. (1991) Enzyme sensor-FIA-system for on-line monitoring of glucose, lactate and glutamine in animal cell cultures. J. Biotechnol., 21: 173-85
27. Cattaneo, M.V., Male, K.B., and Luong, J.H.T. (1992) Chemiluminescence fiber-optic biosensor system for the determination of glutamine in mammalian cell cultures. Biosensors & Bioelectronics 7: 569-574
28. Dremel, B.A.A., Li, S.Y., and Schmid, R.D. (1992) On-line determination of glucose and lactate concentrations in animal cell culture based on fiber optic detection of oxygen in flow-injection analysis. Biosensors & Bioelectronics, 7: 133-139
29. Male, K.B., and Luong, J.H.T. (1992) Determination of urinary glucose by a flow injection analysis amperometric biosensor and ion-exchange chromatography. Appl. Biochem. Biotechnol., 37: 243-254
30. Blum L. J. (1993) Chemiluminescent flow injection analysis of glucose in drinks with a bienzyme fiberoptic biosensor. Enz. Microb. Technol., 15: 407-411
31. Cattaneo, M.V., and Luong J.H.T (1993) On-line chemiluminescent assay using FIA and fiber optics for urinary and blood glucose. Enz. Microb. Technol., 15: 424-428

32. Cattaneo, M.V., and Luong J.H.T (1993) Monitoring glutamine in animal cell cultures using a chemiluminescent fiber optic biosensor. Biotechnol. Bioeng., 41: 659-665
33. Male, K.B., Luong, J.H.T., Tom, R., and Mercille, S. (1993) Novel FIA amperometric biosensor system for the determination of glutamine in cell culture system. Enz. Microb. Technol., 15: 26-32
34. Meyerhoff, E.M., Trojanowicz, K.M., and Palsson, O.B. (1993) Simultaneous enzymatic/electrochemical determination of glucose and L-glutamine in hybridoma media by flow-injection analysis. Biotechnol. Bioeng., 41: 964-969
35. Kyröläinen, M., Hakanson, H., and Mattiasson, B. (1995) On-line calibration of a computerized biosensor system for continuous measurements of glucose and lactate. Biotechnol. Bioeng. 45: 122-128
36. K. Holzhauer-Rieger, W. Zhou, and K. Schügerl (1990) On-line high-performance liquid chromatography for determination of cephalosporin C and by-products in complex fermentation broth. J. Chromatography, 499: 609-615
37. Kurokawa, H., Park, Y.S., Iijima, S., and Kobayashi, T. (1994) Growth characteristics in fed-batch culture of hybridoma cells with control of glucose and glutamine concentration. Biotechnol. Bioeng., 44: 95-103
38. Heinzle, E. (1992) Present and potential applications of mass spectrometry for bioprocess research and control. J. Biotechnol., 25: 81-114
39. Lovrecz, G. and Gray, P. (1994) Use of on-line gas analysis to monitor recombinant mammalian cell cultures. Cytotechnol. 14: 167-175
40. Bonarius, H.P.J., de Gooijer, C.D., Tramper, J., and Schmid, G. (1995) Determination of the respiration quotient in mammalian cell culture in bicarbonate buffered media. Biotechnol. Bioeng., 45: 524-535
41. Backer, M.P., Metzger, L.S., Slaber, P.L., Nevitt, K.L., and Boder, G.B. (1988) Large-scale production of monoclonal antibodies in suspension culture. Biotechnol. Bioeng., 32: 993-1000
42. Behrendt, U., Koch, S., Gooch, D.D., Steegmans, U., and Comer M.J. (1994) Mass spectrometry: A tool for on-line monitoring of animal cell cultures. Cytotechnol. 14: 157-165
43. Oeggerli, A., Eyer, K., and Heinzle, E. (1995) On-line gas analysis in animal cell cultivation: I. Control of dissolved oxygen and pH. Biotechnol. Bioeng., 45: 42-53
44. Eyer, K., Oeggerli, A., and Heinzle, E. (1995) On-line gas analysis in animal cell cultivation: II. Methods for oxygen uptake rate estimation and its application to controlled feeding of glutamine. Biotechnol. Bioeng., 45: 54-62
45. Fleischaker, R.J., and Sinskey, A.J. (1981) Oxygen demand and supply in cell culture. European J. Appl. Microbiol. Biotechnol., 12: 193-197
46. Ramirez, O.T. and Mutharasen, R. (1990) Cell cycle- and growth phase-dependent variations in size distribution, antibody productivity, and oxygen demand in hybridoma cultures. Biotechnol. Bioeng., 36: 839 - 848

47. Kilburn, D.G. (1991) In: Mammalian Cell Biotechnology, M. Bulter, Ed., Oxford University Press
48. Zhou, W., Rehm, J., and Hu, W-S. (1995) High viable cell concentration fed-batch cultures of hybridoma cells through on-line nutrient feeding. Biotechnol. Bioeng., 46: 579-587
49. Kyung, Y.-S., Peshwa, M.V., Gryte, D.M., and Hu, W.-S. (1994) High density culture of mammalian cells with dynamic perfusion based on on-line oxygen uptake rate measurements. Cytotechnol. 14: 183-190
50. Schubert, J., Simutis, R., Dors, M., Havlik, I., and Lübbert, A. (1994) Hybrid modeling of yeast production processes -Combination of a priori knowledge on different levels of sophistication-. Chem. Eng. Technol. 17: 10-20
51. Dors, M., Simutis, R., Lübbert, A. (1995) Hybrid process modeling for advanced process state estimation, prediction and control exemplified at a production scale mammalian cell culture. In Recent Advances in Biosensors, Bioprocess Monitoring, and Bioprocess Control, K.R. Rogers, A. Mulchandani, W. Zhou, Eds., ACS Symposium Series, American Chemical Society, Washington, D.C. (this vol.)

RECEIVED August 14, 1995

Chapter 10

Optical Measurement of Bioprocess and Clinical Analytes Using Lifetime-Based Phase Fluorimetry

Shabbir B. Bambot[1,4], Joseph R. Lakowicz[2], Jeffrey Sipior[2], Gary Carter[3], and Govind Rao[1]

[1]Department of Chemical and Biochemical Engineering and Medical Biotechnology Center of the Maryland Biotechnology Institute, University of Maryland–Baltimore County, Baltimore, MD 21228
[2]Center for Fluorescence Spectroscopy, Department of Biological Chemistry and Medical Biotechnology Center of the Maryland Biotechnology Institute, University of Maryland at Baltimore, Baltimore, MD 21201
[3]Department of Electrical Engineering, University of Maryland–Baltimore County, Baltimore, MD 21228

The measurement of analyte concentrations are a critical part of successful bioreactor and clinical monitoring. While strategies exist for measuring every known analyte, industrial on-line bioreactor control however, is carried out using primarily pH, pO_2 and in some cases cell density measurement and control. This is because the available technology cannot be readily adapted to (a) measure the analyte in an aseptic manner or allow for remote sensing and (b) measure in real time so that on-line control is possible. Similar issues exist for clinical applications. In this review article, we describe a rapidly emerging technology that has the potential of meeting these challenges. It is an optical technique that uses the measurement of fluorescence lifetime rather than intensity for determining the concentration of an analyte.

All the breathtaking advances in genetic and metabolic engineering have not resulted in significant changes in bioreactor or bioprocess monitoring technology. Electrochemical sensors, inspite of various problems, continue to be used on-line for measuring oxygen and pH [1,2] because of a lack of better alternatives. These problems include calibration drifts, flow dependence, slow response times and electrical interferences [1-3]. A variety of other parameters which have been shown to profoundly affect cellular productivities, such as glucose, CO_2, nitrate, phosphate and miscellaneous cation and anion levels in the medium continue to be measured

[4]Current address: AVL Biosense Corporation, 33 Mansell Court, Atlanta, GA 30076

0097–6156/95/0613–0099$12.00/0

off-line due to technological limitations. The long time intervals (typically 10 minutes) required for off-line measurements add to the uncertainties in the measurement as well as to the labor and costs involved. In addition, the need to draw samples frequently contribute to contamination risks making compliance with regulatory requirements difficult. Furthermore, the principles and instrumentation required for each of these measurements varies considerably, thus adding to their cost and complexity. There is, therefore, a strong need for a generic strategy which would permit the real time monitoring of all these parameters. Although the requirements for clinical monitoring are more stringent and regulated, similar concerns prevail. Consequently, although our discussion focuses on application to bioprocess monitoring, key characteristics of these sensors also make them appropriate for measuring clinically important analytes.

Fluorescence detection has been extensively investigated as an alternative to electrochemical sensing.[4-7] Of the various schemes for fluorescence sensing, the most studied, steady state intensity measurements, have not found wide acceptance due to stability and calibration problems. Such measurements are unsuitable for blood or cell culture because of the effects of scattering and absorption by the medium (collectively termed "inner filter effects"). An alternative to intensity measurements is the wavelength-ratiometric method in which (a) the ratio of the intensities at two different emission wavelengths for a single excitation wavelength or (b) a ratio of the intensities obtained at a single emission wavelength for two different excitation wavelengths is plotted as a function of analyte concentration. Wavelength-ratiometric methods are independent of probe concentration and inner filter effects but have not been applied extensively because of the lack of intensity-ratio probes.

Fluorescence Lifetime Measurement

The fluorescence lifetime (also referred to as time resolved fluorescence) represents the average amount of time a fluorophore spends, following light absorption, in the excited state before returning to the ground state. Lifetime measurement is therefore, distinct from intensity based methods which rely on the measurement of steady state fluorescence. For lifetime measurement applications fluorophores are chosen or designed for their change in fluorescence lifetime upon interacting with the specific analyte they are intended to sense. Of the two available methods for measuring fluorescence lifetimes, time-domain and frequency-domain, the former calls for expensive instrumentation and complex electronics. Fortunately, recent developments in the frequency-domain technique known as *phase-modulation fluorimetry*, have made possible reliable and inexpensive measurements of nanosecond lifetimes [8]. In phase-modulation fluorimetry, sinusoidally modulated excitation light excites a fluorescent chemical probe. The fluorescence emission is, as a result forced to oscillate at the same frequency. Because of the analyte dependent finite fluorescence lifetime of the probe, the emission is phase shifted and demodulated with respect to the excitation. Both phase and demodulation values provide independent measurements of the analyte concentration, adding to the robustness of the technique.

With reference to figure 1(a), the phase shift (Φ) and the demodulation factor (m) are related to the fluorescence lifetime τ of single exponentially decaying fluorophores according to the relations :

$$\tan \Phi = \omega\tau \quad \text{and} \quad m = [1 + \omega^2\tau^2]^{-1/2} \qquad \text{Equation (1)}$$

where ω is the frequency of modulation. While single exponentially decaying fluorophores are rarely encountered in practice, direct calibrations between the analyte concentration and a measured parameter e.g. Φ are readily established using calibration methods.[9]

Decay time sensing is advantageous because such measurements are independent of probe concentration, photobleaching and drifts in lamp intensity, are independent of inner filter effects, and do not require dual emission from the probe, all of which, are major limitations of current fluorescence intensity based optrodes. The recent availability of novel synthetic fluorophores, particularly those with long lifetimes (in hundreds of nanoseconds) has further facilitated the early transfer of this powerful technique into practical applications. It is conceivable that in a few years, bioreactor monitoring will be primarily implemented through optical means. Although, we have, in this review, primarily focused upon *in situ* pO_2, pH and pCO_2 measurements, the methodology is generic and can be readily extended to measure a variety of other analytes.

Oxygen Sensor

Oxygen holds great significance as an indicator of the metabolic state of living cells. Consequently much effort in bioreactor design and operation is directed toward the monitoring and control of oxygen supply to the fermentor. Despite its many limitations the amperometric Clark electrode[10] continues to be widely used for this purpose. This is a consumption based sensor where oxygen (and electrolyte) is consumed whilst being sensed. The sensor suffers from flow dependence, slow response times, calibration drifts and electrical interference. Alternative technologies in this area are long overdue. It is now possible to monitor oxygen tensions through an optical alternative that is comparable in cost and superior in performance to the Clark electrode.

Figure 1(b) illustrates the principles and instrumentation involved in such an optical oxygen sensor. We note that phase angle sensing of oxygen was first described by Wolfbeis and co-workers.[11] The excitation light is provided by an inexpensive blue LED intrinsically modulated at 76 kHz. This low frequency modulation, permitted by the long lifetime of the synthetic fluorophore makes it possible to use simple radio type electronics. The phase fluorimetric oxygen sensor measures the quenching by oxygen of the transition metal complex, tris [4,7-diphenyl-1,10-phenanthroline] ruthenium $(II)^{2+}$.[12] This transduction scheme generally follows Stern Volmer kinetics [8] which predict a first order type response to oxygen tension resulting in significantly higher oxygen sensitivities at low

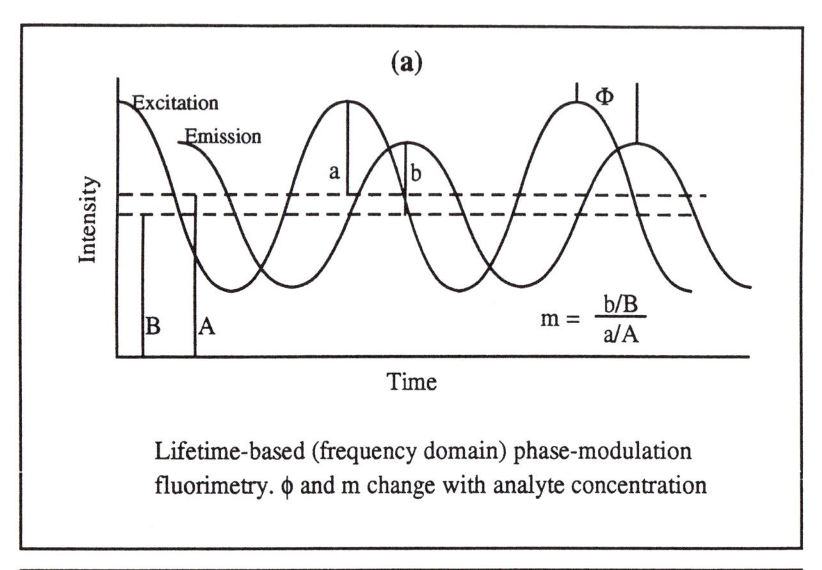

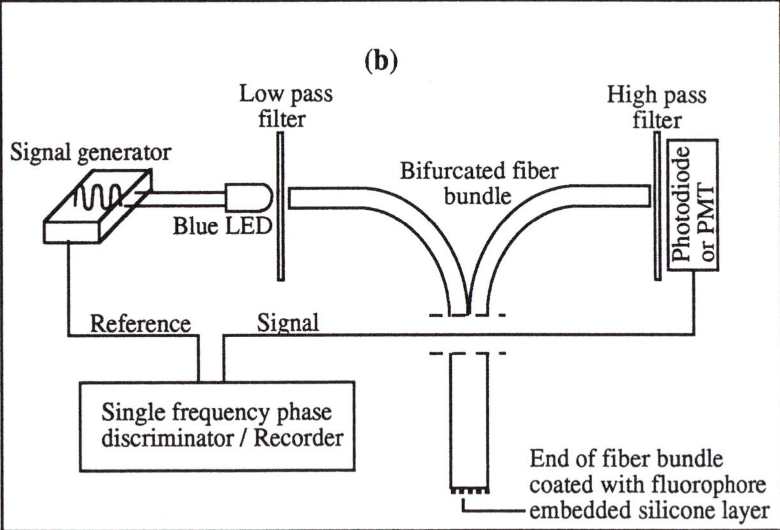

Figure 1. (a) Fluorescence lifetime measurement using phase-modultion fluorimetry. (b) Schematic of phase fluorimetric oxygen sensor.

oxygen concentrations[9] as compared with the Clark electrode. The fluorophore is immobilized in a silicone rubber membrane which is an ideal matrix for this application because of its high oxygen solubility and diffusivity. The sensor is equally sensitive to gaseous and dissolved oxygen tensions. The silicone matrix selects for gaseous analytes because of its hydrophobic nature, adding to sensor specificity. In addition, the hydrophobic matrix and the water insolubility of the ruthenium complex add to long term sensor stability by preventing probe wash out and leaching. The sensor performed satisfactorily in a bioreactor environment, was autoclavable and was found to be superior to current oxygen monitoring technology viz. the Clark electrode in that it is maintenance free, shows no drift in calibration over weeks of operation and has a faster response.[9]

The recent availability of Os complexes which have absorbance maxima that are red shifted (greater than 650 nm) has generated the possibility of non-invasive oxygen sensing, given that the human skin as well as most biological tissue does not absorb strongly and have significantly reduced auto-fluorescence when illuminated with light at this wavelength.[13] The Clinical assessment of oxygenation in arteries and specific tissue is of significant importance in monitoring critically ill patients. Arterial oxygenation problems are a common indication of loss of hemoglobin or deteriorating lung function. Similarly low tissue oxygen indicates tissue hypoxia and loss of function. In general, disturbances in oxygenation result in decreased performance from organs highly dependent on oxygen such as the heart and brain.

We have demonstrated phase angle sensing of oxygen (Bambot, S. B. *et al. Biosensors and Bioelectronics.,* In press.), through skin, using the complex [$Os(2,2',2''\text{-terpyridine})_2^{2+}$] (reference 14) encouraging the possibility of low-cost transcutaneous sensing of tissue oxygen. Experiments were conducted using layers of chicken skin to block both excitation and emission light paths during spectra acquisition. The sample was a saturated solution of the osmium complex in methanol held in a cuvette. The excitation light source was a 635 nm laser diode (Toshiba, model TOLD9520, Edison, NJ) intrinsically modulated at 750 kHz. Chicken skin was used because of its easy availability and its optical similarity to human skin.[15] Figure 2 shows real time through skin fluorescence intensity and phase angle measurements as oxygen and nitrogen are alternatingly bubbled through the fluorophore solution. The measurements were made first with no skin, one layer of skin and finally two layers of skin in both excitation and emission paths. While a 50 fold decrease in the intensity response occurred the phase angle measurements were unaffected by the presence of the skin. The experiment demonstrates the independence of phase angle measurements to changes in skin thickness and attenuation in excitation light intensity. Provided that a minimum measurable emission is available (the quantity of light required for measurement at a tolerable signal to noise level, is governed entirely by the photodetector) the phase angles are independent of the attenuating effects that limit fluorescence intensity based measurements.

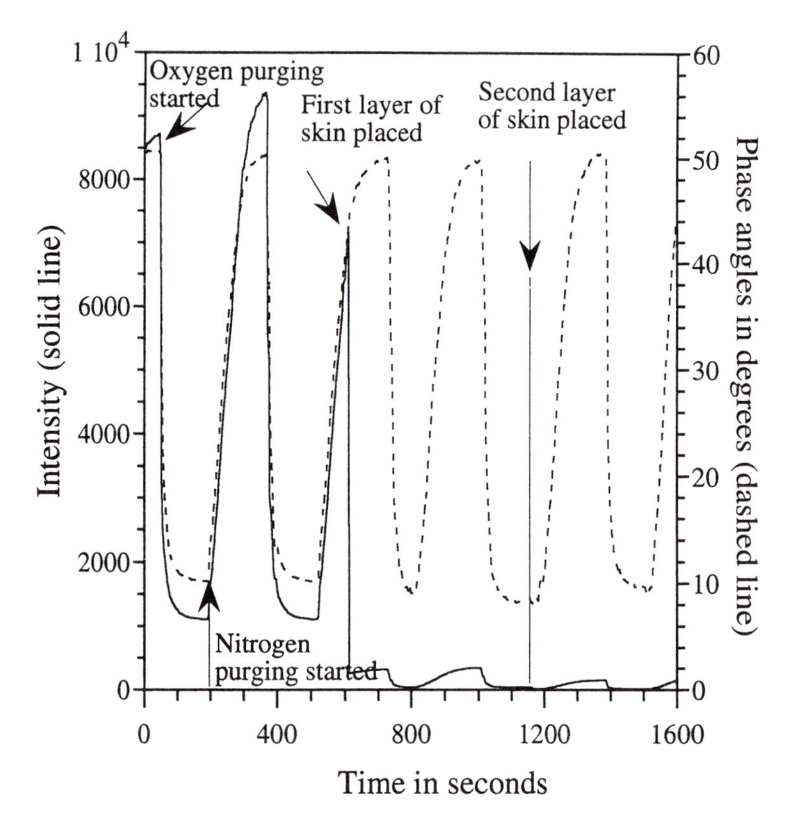

Figure 2. Oxygen dependent changes in fluorescence lifetime and intensity of Os(2,2',2"-terpyridine)$_2^{2+}$ as observed through layers of chicken skin placed in the excitation and emission paths.

pH/pCO_2 sensor

Real time pH monitoring is necessary for maintaining cell viability and production and the pH glass electrode is widely used for this purpose. pH electrodes however, are far from perfect. In process monitoring, it is rarely possible to measure pH with an accuracy better than 0.25 pH units.[2] Ground loop currents, interactions between sample and electrolyte solutions, acid and alkali errors and frequent servicing requirements are major limitations of these sensors. The measurement of pCO_2 is essentially that of pH and is often used in determining the metabolic state of the cells in a fermentation and in evaluating the oxygen uptake rate (OUR). A pH electrode in contact with a bicarbonate solution and isolated from the sample fluid by a Teflon or plastic membrane is routinely used for this purpose.[16] CO_2 diffuses in (or out) through the membrane and changes the bicarbonate solution pH in proportion to the CO_2 partial pressure in the sample. In addition to all the drawbacks of a pH glass electrode, the requirement for water in such a system (bicarbonate solution) results in a slow response and is understandably bulky. This perhaps explains why it is not widely used in biochemical engineering laboratories and process streams despite its considerable metabolic significance.[17]

Optical pH sensing using phase modulation fluorimetry has been demonstrated using fluorescent pH sensitive dyes.[18] However, the dyes have nanosecond lifetimes requiring high modulation frequencies (>100 MHz) and blue-green light (450- 550 nm) excitation translating once again into high costs that prevent routine applications. In order to develop inexpensive lifetime-based pH sensors, an alternative transduction strategy compatible with inexpensive red light sources such as red laser diodes (> 635 nm) that permit MHz modulation was required.

Such a strategy was made possible by employing the principle of Fluorescence Resonance Energy Transfer (FRET).[19] By using FRET (figure 3), the requirements usually made of a single fluorophore (*viz.* high quantum yield, red excitable fluorescence and sensitivity to a specific analyte) are now distributed over two molecules. For pH sensing, a high quantum yield, pH insensitive, fluorescence *donor* (texas red hydrazide) and a non fluorescent, pH sensitive, *acceptor* whose absorbance overlaps significantly with the donor's emission is used. FRET is a non-radiative distance-dependent dipole dipole type interaction in which the acceptor quenches the donor fluorescence.[8,20] Optimal energy transfer is obtained when the donor acceptor pair are within 40-70 Å (known as the Förster distance of energy transfer) of each other. The degree of quenching is governed by the quantum yield of the donor, the extinction coefficient of the acceptor, the spectral overlap between the emission of the donor and the absorbance of the acceptor and of course, the spatial distance between the two molecules. This distance was ensured by co-immobilizing donor and acceptor molecules at high concentrations in a hydrophilic "proton permeable" sol-gel matrix.[21]

Sol-gel is a room temperature glass forming technology [22] that results in a unique water permeable matrix with good optical properties. Such matrices have

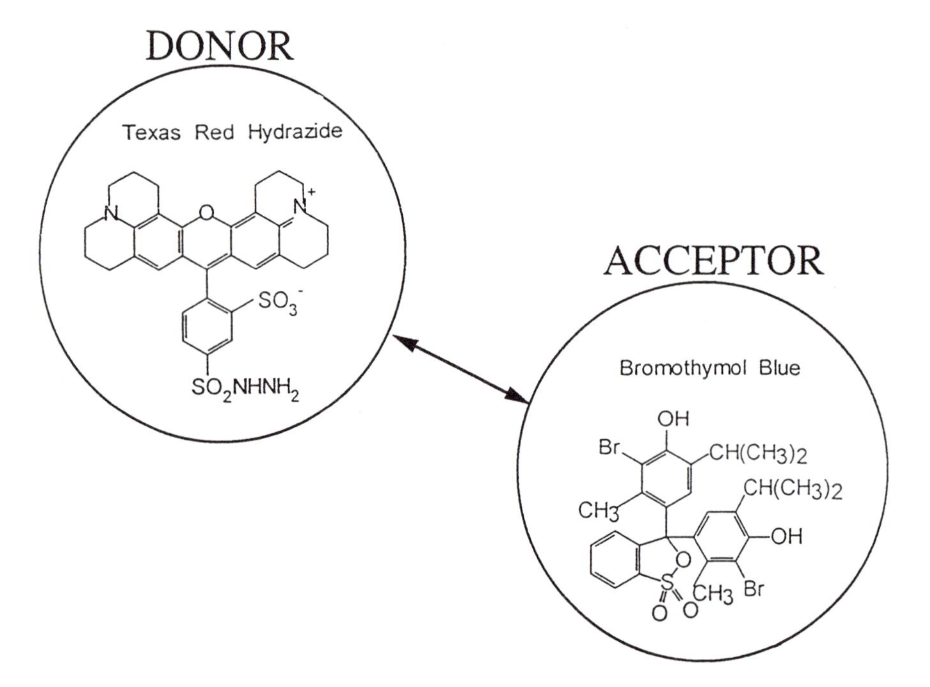

Figure 3. Fluorescence Resonance Energy Transfer in pH sensing. The energy transfer is dependent upon the distance between donor and acceptor molecules, the extent of spectral overlap between donor emission and acceptor absorbance, the quantum yield of the donor and extinction coefficient of the acceptor.

previously been used to entrap organic dyes (for review see reference 23) and have been found to provide an environment that is optically and chemically inert. Using Texas Red Hydrazide as the donor and Bromothymol blue as the acceptor, entrapment of the dyes in the Sol-Gel matrix resulted in a unique, chemically active, optical sensing material.[21] The sensor was compatible with red light sources and by using thin coatings of this matrix over normal glass substrates fast (millisecond) response times were observed. The pH dependent changes in fluorescence intensity and phase angles were reproducible and reversible and the sensor was found to be stable and retain its pH sensing capabilities even after repeated autoclaving.[21]

The principles described above in developing a phase fluorimetric pH sensor were directly applicable to CO_2 sensing. In a recent report we have described the development of such a sensor (Sipior J., *et al. Analytical Biochemistry.*, In press.) using the immobilization strategy described by Mills et al.[24] and our pH sensitive FRET dyes. In this sensor all the necessary ingredients, including water molecules were incorporated in a hydrophobic (plastic) film with the help of a phase transfer catalyst.[24,25] The CO_2 diffusing into the plastic matrix results in the generation of protons that interact with the pH sensitive energy transfer dyes. Both sensors (that of Mills et al.[24] and ours) were found to be sensitive to CO_2 and like the oxygen sensor maximum sensitivity was observed at low (below 5%) CO_2 concentrations. The sensors demonstrated fast (seconds) response times and a long shelf life. However prolonged use in both sensors led to the permanent generation of the protonated form of the dye over a period of 20-100 hours.

Conclusions

Advances in fiber optics, light sources, detectors and other instrumentation (largely due to the communications industry) coupled with the availability of novel synthetic fluorophores have made practical applications of this technology possible and will continue to do so. The generic nature of this technology will inevitably find applications in other areas (Environmental, toxicological and chemical process monitoring to name a few). Importantly, bioprocess monitoring also serves as a proving ground for the ultimate use of these sensors in clinical monitoring and diagnostics. One example application where lifetime sensing can provide a significant improvement is flow cytometry [13]. Currently, this technique is used for cell sorting based on large fluorescence intensity differences between these cells such as those resulting from cell division or the presence of cell surface antigens. Large intensity changes are required because there are substantial cell by cell variations due to differences in cell size and shape. By using phase fluorimetry, one can further "fine tune" flow cytometry and can extend it to differentiate cells based on differences in intracellular physiology.[13]

Judicious component selection, low cost light sources and simple electronics, will help in further reducing instrument cost as will the use of multiple analyte sensing with one instrument. Coupled with advances in chemistry, it is conceivable that the next generation of bioprocess and clinical sensors will rapidly move into the optical realm. The field of lifetime based sensing has advanced dramatically during the past year. In addition to the probes for O_2, pH and pCO_2 new lifetime probes have become available. We are now aware of a wide variety of probes which change lifetime in response to glucose, Ca^{2+} [26,27], Mg^{2+},[28], Na^{+},[28], K^{+},[29] and Cl^{-} ,[29]. The generic technology described can be directly ported over upon characterization of the new probes and development of effective immobilization methods.

Acknowledgements

This work was supported by grants (BCS 9157852, BCS-9209157 and BES-9413262) from the National Science Foundation, with support for instrumentation from the National Institutes of Health grants RR-08119 and RR-07510. Matching funds were provided by Artisan Industries, Inc., and Genentech, Inc.

Literature Cited.

1. Lee, Y-H and G. Tsao.; In *Advances in Biochemical Engineering;* Editors, T.K. Ghose, A. Fiechter, and N. Blakebrough; Springer-Verlag: Berlin, 1979, Volume 13, pp. 35-86.
2. McMillan, G. K.; *Chem. Eng. Prog.* **1991,** *87,* 30-37.
3. Buehler H. W., Bucher. R.; In *Sensors in Bioprocess Control* Editors, J. V.,Twork and A. M.,Yacynych; Marcel Dekker: New York, 1990, pp. 127-171.
4. *Fiber-Optic Chemical Sensors and Biosensors.;*Wolfbeis, O. S., Ed.; CRC Press: Boca Raton, 1991; Vols. I and II.
5. Seitz, W. R.; *Anal. Chem..* **1984,** *56 (1),* 18A-34A.
6. Peterson J. I., Vurek, G. G; *Science.* **1984,** *224,* 123-127.
7. Angel, S. M.; *Spectroscopy..* **1987,** *2(4),* 38-48.
8. Lakowicz, J. R. *Principles of Fluorescence Spectroscopy.* Plenum Press: New York. 1983.
9. Bambot, S. B., R. Holavanahali, J. R. Lakowicz, G. M. Carter and G. Rao.;*Biotechnol. Bioeng.* **1994,** *43,* 1139-1145.
10. Clark, Jr., L. C.;*Trans. Am. Soc. Artif. Int. Organs.* **1956,** *2,* 41-54.
11. Lippitsch, M. E., J. Pusterhofer, M. J. P. Leiner and O. S. Wolfbeis.; *Anal. Chim. Acta.* **1988,** *205,* 1-6.
12. Lin, C-T., Böttcher, W., Chou, M., Creutz, C. and Sutin, N.; *J. Am. Chem. Soc.* **1976,** 98, 6536-6544.
13. Lakowicz, J. R., *Laser Focus World.* **1992,** *28(5),* 60-80.
14. Kober, E. M., Marshall, J. L., Dressnick, W. J., Sullivan, B. P., Caspar, J. V. and Meyer, T. J.; *Inorg. Chem..* **1985,** *24,* 2755-2763.
15. Anderson, R. R. and Parrish, J. A. In *The Science of Photomedicine,* Editor J. D. Regan and J. A. Parrish.; Plenum press: New York, 1982, pp. 149-194.
16. Severinghaus, J. W. and Bradley, A. F., J.; *Appl. Phys.* **1958,** *13,* 515-520.
17. Gommers, P. J. F., B. J. van Schie, J. P. van Dijken and J. G. Kuenen, (1988) *Biotechnol. Bioeng.* **32,** 86-94.
18. Szmacinski, H. and Lakowicz, J. R.; *Anal. Chem..* **1993,** *65,* 1668-1674.
19. Lakowicz, J. R., H. Szmacinski and M. Karakelle. *Anal. Chim. Acta.* **1993,** *272,* 179-186.

20. Förster, Th., Translated by R.S. Knox.; *Ann. Phys. (Liepzig).* **1948,** *2,* 55-75.
21. Bambot, S. B., Sipior, J., Lakowicz, J.R. and Rao, G., *Sensors & Actuators B.* **1994,** *22,* 181-188
22. Brinker, C. J. and Scherer, G. *Sol-Gel Science: The Physics and Chemistry of Sol-Gel Processing.;* Academic Press: Boston. 1990.
23. Avnir, D., Braun, S. and Ottolenghi, M.; *National Technical Information Service (NTIS) Publication No. AD-A224.* **1991,** *154,* 21 pp.
24. Mills, A., Chang, Q. and McMurray, N; *Anal. Chem.* **1992,** *64,* 1383-1389.
25. Mills, A. and Chang, Q.; *Int. Pat. Appl.* **1993,** WO 93/14399.
26. Lakowicz, J. R., Szmacinski, H., Nowaczyk, K. and Johnson, M. L.; *Cell Calcium.* **1992,** *13,* 131-147.
27. Szmacinski, H., Gryczynski, I. and Lakowicz, J. R.; *Photochemistry and Photobiology.* **1993,** *58,* 341-345
28. Szmacinski, H. and Lakowicz, J. R.; In *Probe Design and Chemical Sensing, Topics in Fluorescence Spectroscopy;* Editor, J. R. Lakowicz.; Plenum Publishing Corporation: New York, 1994, Vol. 4, pp. 295-334.
29. Lakowicz, J. R., Szmacinski, H.; *Sensors and Actuators B.* **1993,** *11,* 133-143.

RECEIVED June 5, 1995

Chapter 11

Biosensor for On-Line Monitoring of Penicillin During Its Production by Fermentation

Canh Tran-Minh and Helmut Meier

Ecole Nationale Supérieure des Mines, Centre Science des Processus Industriels et Naturels/Biotechnology, 158, Cours Fauriel, 42023 Saint Etienne Cedex 2, France

An enzyme sensor has been developed for the on-line monitoring of penicillin V during its production by fermentation. The enzyme is crosslinked as a very fine film to the sensitive end of a pH transducer, using the spraying method. The biosensor is incorporated in a flow injection analysis (FIA) system within a home-made stirred flow detection cell. This new and improved biosensor shows good stability and large concentration measuring range. Penicillin-V in fermentation broth is detected during the whole fermentation process and the results are compared with the HPLC method. On-line measurements are achieved through the automation of the FIA system.

In the field of chemical analysis, biosensors have undergone rapid development over the last few years. This is due to the combination of new bioreceptors with the ever-growing number of transducers [1]. The characteristics of these biosensors have been improved, and their increased reliability has yielded new applications. Recently, a new technique of enzyme immobilization has been developed to obtain biosensors for the determination of enzyme substrates [2]. It is based on the enzyme adsorption followed by a crosslinking procedure. Therefore, a penicillin biosensor can be obtained and associated with a flow injection analysis (FIA) system for the on-line monitoring of penicillin during its production by fermentation [3-4]. This real-time monitoring of bioprocess would lead to optimization of the procedure, the yield of which could then be increased and the material cost decreased.

Experimental

Enzyme Electrode. The determination of penicillin is possible by immobilising penicillinase onto glass electrodes. The principle of operation of the sensor is based on the changes in the H^+ concentration resulting from the enzymatic hydrolysis of penicillin by penicillinase .

Several methods for enzyme immobilization can be found in literature. In our laboratory, we have developed a new enzyme immobilization technique making possible response time of the biosensor much inferior to any of the response times so far reported for a penicillin sensor: the combined pH electrode to be coated with the enzyme was left

0097–6156/95/0613–0110$12.00/0

overnight in sodium phosphate buffer at room temperature (used electrodes were dipped in 1M NaOH and 1M HCl, half an hour, each 3-4 times, alternatively, prior to its standing overnight in the buffer).

The electrode was then wiped dry with lens paper before being immersed for 15 minutes in an enzyme solution, containing 4 $mg.ml^{-1}$ of penicillinase. After drying the electrode for 20 min at 4 °C, it was mounted on a rotator horizontally, as shown in Figure 1.

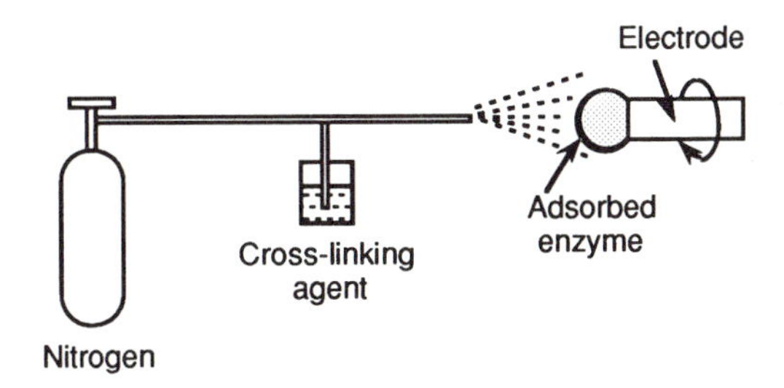

Fig 1. Deposition of thin enzymatic membranes in the construction of glass enzyme electrodes.

A solution containing 2.5 % glutaraldehyde in distilled water was then sprayed over the sensitive end of the electrode using an air-brush, under a pressure of 1.5 bar, nitrogen gas, at room temperature, keeping the electrode in rotation at 50 rpm. The diaphragm of the reference electrodes was covered by plastic clips to prevent a contact with the enzyme and glutaraldehyde solution during the immobilization period.

The reticulation between the enzyme and the glutaraldehyde over the electrode was allowed to continue for 15 min at 4 °C, before rinsing it in the buffer at room temperature for 5 minutes. The electrode was stored in the same buffer at 4 °C, when not in use.

Detection Cell. The configuration is shown in Figure 2. Penicillinase was immobilized over a combined pH-glass electrode: type LoT 403-M8-S7, Ingold, Paris. A home-made stirred flow cell is used as detection chamber. PVC and PTFE were used to construct the cell; small stainless steel tubes were used as fluid inlets (in order to connect the tubes to the cell). The potentiometric measurements were obtained using a Radiometer pHM-64 research pH meter connected to a Sefram-Recorder (Sefram-Servotrace, Paris) and to the computer. The detection cell was agitated by a stirrer (Microlab, Aarhus, Denmark) at a moderate speed.

Sample and buffer solutions were pumped by two 2-channel peristaltic pumps (Ismatec SA, Zürich, Switzerland). using tygon tubes 0.51, 1.42 and 1.52 mm inner diameter (Bioblock, Illkirch, France).

Automation Procedure. Automatic sample injection is carried out by use of a timing control pneumatic actuator connected with a 4-way teflon injection valve (both Rheodyne, Cotati, U.S.A.). The actuator permits automatic operation of the valve.

The analog millivolt signal arising from the pH-meter is converted to digital equivalent by a 6B11 module mounted on a 6BP04-2 backplane. The digital I/O backplane is a 6B50-1 module which provides 24 digital I/O channels. These backplanes are from Analog Devices, Norwood/U.S.A. The I/O channels are connected to solid-state relays plugged in a EGS08000 backplane from Celduc, Sorbiers/France. These relays control the injection valve and the pumps.

A MacIntosh computer is connected to the 6BPO4 backplane via the serial port (RS-232).

The software is written in Quick Basic, and it is structured, with subroutines for control of the injection valve and the pumps, data aquisition, calibration, and calculations.

Subroutines ensure recording of the baseline, injection of sample, recording of response signal and preparation of the next sample injection. The record signals are saved and by means of calculation and calibration subroutine, the mV-peak height is converted to the appropriate concentration of the substrate. Four calibration samples with different concentrations are used to trace the calibration curve.

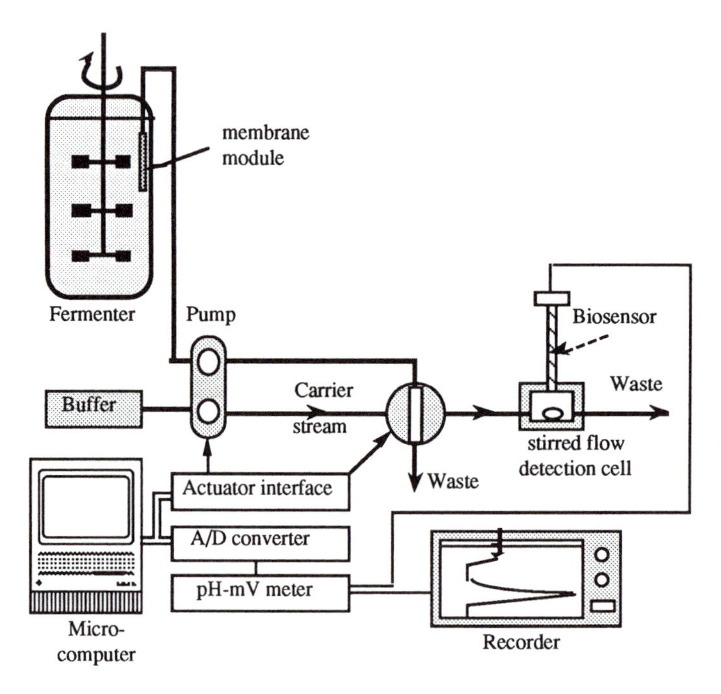

Fig. 2: Automated Biosensor-FIA system for penicillin monitoring

Flow Injection System. The manifold is shown in Figure 2. The baseline was obtained by pumping a buffer solution through the detection cell (pH meter in millivolt mode). The sample flow rate was always 0.78 $ml.min^{-1}$. Penicillin standard solutions of 0.1 mM to 65 mM were prepared in the fermentation broth. The diluted samples were injected into the carrier by means of the injection valve which is placed close to the detection cells in order to minimise the delay between injection and detection. The dead volume chosen was 2 ml. The cells are thermostatted (25°C). For measurements, the potential difference between the peak height and the base line was automatically recorded by the computer and the recorder. The potential of the electrode always returned to its base line as soon as a fresh buffer solution come again to the contact of the electrode.

Penicillin in fermentation broth were measured after sample withdrawal from the bioreactor. The samples were filtered by an *in situ* membrane module with a 0.22 µm nominal pore diameter.

Results and Discussion

The necessity to measure penicillin in fermentation broth is important. Penicillin fermentations typically run for 10 days and a high frequency of analysis is therefore not necessary. However, this determination is done within a large concentration range and in a complex media. During the fermentation process, there is a significant change in the medium composition, whereas the pH value of the broth remains approximately constant. To fulfil these requirements, injection volume, flow rate and buffer strengths were adjusted to have good reproducibility as well as acceptable sensitivity and response time.

The penicillin calibration curve, conducted before the beginning of the fermentation monitoring, is shown in Figure 3. It can be seen that there is a large linear measuring range: the maximum penicillin concentration at the end of the fermentation is about 60 mM. This fact presents an important advantage in view of simplicity of the fermentation monitoring because one single calibration curve covers the whole concentration range.

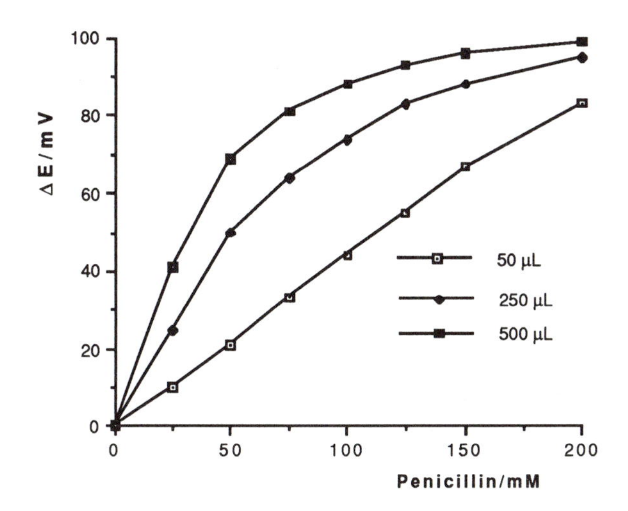

Fig. 3: Calibration curve for the penicillin sensor as a function of sample injection volumes: (50, 250, 500 µL)

The comparison of the results of the total penicillin V concentration versus the fermentation time, obtained by the reference method (HPLC) and by the biosensor-FIA system is shown in Figure 4 for the fermentation FBO-23. Penicillin-V together with a small concentration range of para-hydroxy-penicillin-V are produced during the fermentation. Both penicillin V possess a ß-lactam ring and are detected by the biosensor. The Figure shows good agreement between the two off-line measurements over a wide range of concentration with a better repeatability of the biosensor measurements. The operating conditions chosen for FIA allow measurements without any change in parameters during the whole fermentation process. The pH of the broth is regulated to about 6.5 ± 0.2. The Figure shows good agreement of the biosensor responses over a wide range of concentration with the HPLC measurements.

Stability of the biosensor. The long term stability of the two electrodes was examined at room temperature. After construction, 2 penicillinase electrodes were used in the FIA system for about 8 hours per day. Between measurements, the sensors were stored in the detection cells, always at room-temperature. The penicillinase electrode shows a good stability (98% of initial response) during 33 days (Fig. 5).

Response time. The very thin film of immobilized enzyme allows rapid diffusion of substrate and product which results in an almost instantaneous state of equilibrium. Steady-state response is achieved in less than 10 seconds. This response time has been compared to the response times reported by other investigators as from 1 to 20 minutes [5].

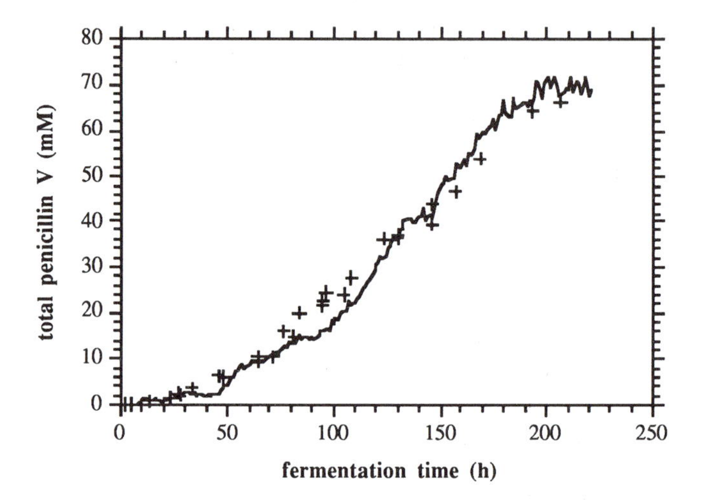

Fig. 4: Determination of penicillin V during the fermentation process with a biosensor (—) and HPLC (+) methods

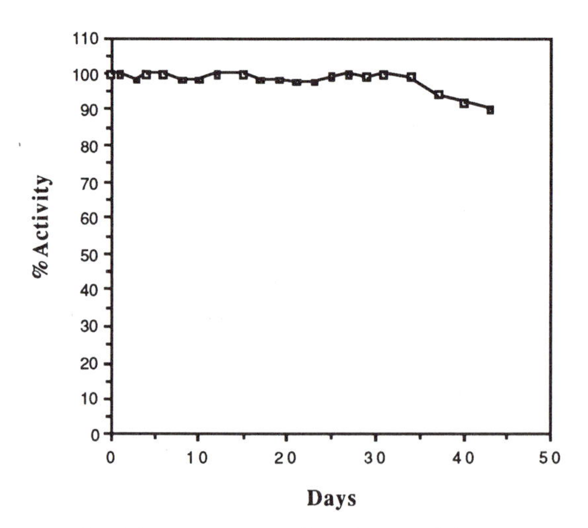

Fig. 5: Long term stability of the penicillinase sensor
Penicillin V concentration in fermentation broth: 15 mM, pH: 6.5

Conclusion

Flow injection analysis (FIA) combined with biosensors offer new applications in the field of automated analysis and process control. Biosensors with rapid response time ensure high sample throughput, improve the sensitivity and decrease the cost of analysis.

Using a new immobilisation technique, it is possible to obtain a fast responding enzyme sensor using an ordinary pH glass electrode. This type of sensor can provide substrate concentration after only a few seconds. Enzyme electrodes are directly incorporated in a home-made stirred flow detection cell. Automation of the biosensor-FIA system make possible on-line monitoring of penicillin-V during its production by fermentation. The various conditions required for penicillin determination during the fermentation process are reduced or completely eliminated by the sample dilution, which takes place in the detection cell. Investigations concerning the optimal operating conditions of the FIA system have also revealed so that the operating parameters can be maintained over the whole penicillin concentration range.

The present biosensor/FIA system is simple, reliable and inexpensive to determine penicillin in fermentation broth samples and it can also be used for the determination of other biological species [6].

Literature Cited

1. Tran-Minh, C. *Biosensors*; Chapman & Hall: London, 1993.
2. Kumaran, S.; Meier, H.; Danna, A. M.; Tran-Minh, C. *Anal. Chem.* **1991**, *63*, 1914-1918.
3. Nielsen, J.; Nikolajsen, K.; Villadsen, *J. Biotechnol. and Bioeng.* **1989**, *33*, 1127-1134.
4. Meier, H.; Tran Minh C. *Anal. Chim. Acta* **1992**, *264* (1), 13-22.
5. Meier, H.; Kumaran, S.; Danna, A. M.; Tran-Minh, C. *Anal. Chim. Acta* **1991**, *249*, 405-411.
6. Meier, H.; Lantreibecq, F.; Tran-Minh, C. *J. Automatic Chem.* **1992**, *14* (4), 137-143.

RECEIVED August 28, 1995

Chapter 12

Selective Measurement of Glutamine and Asparagine in Aqueous Media by Near-Infrared Spectroscopy

Xiangji Zhou[1], Hoeil Chung[1,3], Mark A. Arnold[1,4], Martin Rhiel[2], and David W. Murhammer[2]

[1]Department of Chemistry and [2]Department of Chemical and Biochemical Engineering, University of Iowa, Iowa City, IA 52242

Asparagine and glutamine concentrations have been determined in binary aqueous solutions with near-infrared (NIR) absorption spectroscopy. Spectra were collected over a range from 5000 to 4000 cm^{-1} with a 1.0 mm optical path length cell. Differences in absorbance features around 4570 and 4380 cm^{-1} for asparagine and glutamine provide the analytical information required for the resolution of these similar amino acids. The best overall performance was obtained by partial least-square (PLS) regression coupled with digital Fourier filtering over the spectral range of 4800-4250 cm^{-1} for asparagine and 4650-4320 cm^{-1} for glutamine. Asparagine measurements were possible over the concentration range from 1.0 to 11.7 mM with a standard error of prediction (SEP) of 0.18 mM and a mean percent error of 2.50%. Glutamine could be measured over the concentration range from 1.0 to 13.6 mM with a SEP of 0.10 mM and a mean percent error of 2.00%. These results represent a critical first step in developing a NIR spectroscopic method for monitoring asparagine and glutamine in mammalian and insect cell cultures.

Monitoring and control of processes are becoming increasingly important in the agricultural, pharmaceutical, textile, food and other industries (1, 16). For animal cell cultures, it is necessary to properly control the feeding of nutrients, removal of products and accumulation of by-product inhibitors in order to increase efficiency and reduce the cost of production (2-6).

Enzyme-based biosensors (1), and NIR spectroscopy (7,10) are currently being developed as continuous monitors for bioreactors. Enzyme-based biosensors are

[3]Current address: Chemical Process Research Lab, Yukong Limited, Ulsan, Korea
[4]Corresponding author

0097–6156/95/0613–0116$12.00/0

not ideal for monitoring bioreactors. Although biosensors are relatively sensitive and can be selective for critical bioreactor analytes, such as glucose and amino acids, they are invasive in nature and are severely limited in terms of operational lifetime under conditions of most bioreactors. In addition, enzyme-based biosensors are difficult to sterilize and demand periodic calibration.

In contrast, remote NIR spectroscopy is noninvasive and can, in principle, operate for extended periods of time without recalibration. A NIR approach is also capable of extracting quantitative information for many components from a single spectrum (7). The primary concern for NIR spectroscopy is the ability to accurately differentiate similar compounds given the rather broad and overlapping nature of NIR absorbance bands.

Biologically important species, including amino acids, have unique spectral features which can be exploited to provide measurement selectivity. In the region of 5000 - 4000 cm^{-1}, the major absorption contributions come from combinations of vibrational transitions for aliphatic C-H, alkene C-H, amine N-H (ionized or not) and O-H bonds (8). Such spectral differences have been used to selectively measure structurally different compounds, such as glucose and ammonium ions (9), and glucose and glutamine (10).

We have performed a study designed to establish the utility of NIR spectroscopy for distinguishing chemically similar compounds. Our test system is a set of binary mixtures of glutamine and asparagine in an aqueous buffer solution. These amino acids differ by a single methylene group in the side chain. The additional methylene group present in glutamine affects the spectral absorbance feature around 4400 cm^{-1} which is predominately composed of combination bands of stretching and bending transitions of the C-H bonds (11). Individual absorbance spectra are presented in Figure 1 for asparagine and glutamine. Comparison of these spectra reveals small differences in the 4400 cm^{-1} region. Our investigation has established that these small spectral differences are sufficient to differentiate glutamine and asparagine at the millimolar concentration level. PLS regression coupled to digital Fourier filtering has been used for these measurements.

Experimental

Apparatus, Reagents and Solutions. A Nicolet 740 Fourier transform infrared (FTIR) spectrometer (Nicolet Analytical Instruments, Madison, WI) was used for the collection of all spectra. A 250 W tungsten-halogen lamp, a calcium fluoride beamspilitter and a cryogenically cooled indium antimonide (InSb) detector were used. A multilayer optical interference filter (Barr Associates, Westford, MA) was used to confine the spectral range of 5000 - 4000 cm^{-1}. The sample cell was a rectangular infrasil quartz cell with a path length of 1 mm (Wilmad Glass Co., Buena, NJ). Temperature was controlled by placing the sample cell in an aluminum-jacketed cell holder in conjunction with a VWR 1140 refrigerated temperature bath (VWR Scientific, Chicago, IL). Temperatures were measured with a copper-constantan thermocouple probe placed directly in the sample solution and reading were obtained from an Omega Model 670 digital meter (Omega Inc., Stamford, CT).

Asparagine and glutamine were purchased from Sigma Chemical Co. (St. Louis, MO). Sodium bicarbonate and sodium phosphate monohydrate were obtained from Aldrich Chemical Company, Inc. (Milwaukee, WI). All aqueous solutions were prepared in deionized water purified with a Milli-Q Reagent Water System (Millipore, Bedford, MA).

Sixty-six different binary mixtures were prepared by randomly mixing different milligram quantities of asparagine and glutamine in 10 ml of buffers. The buffer was composed of 4.17 mM bicarbonate and 8.44 mM phosphate, and the pH was set at 6.35. All measurements for a particular solution were completed in less than one hour after the solution was prepared. Fresh samples were used to avoid complications caused by hydrolysis of these amino acids.

Spectra Collection and Processing. Spectra collection started after placing the sample in the sample holder and waiting for a two minute equilibration period. Spectra were collected as double-sided interferograms of 256 co-added scans. After triangularly apodized, the interferograms were Fourier transformed to generate single beam spectra with a spectral resolution of 1.9 cm^{-1}. Single beam spectra were later transferred to an Iris Indigo computer (Silicon Graphics, Inc., Mountain View, CA) for processing. Three single spectra were collected sequentially for each sample solution without moving the sample. Blank buffer solutions were used for the collection of background single beam spectra. After every fourth sample measurement, a new background spectrum was collected and used as the reference spectrum for subsequent sample spectra. Each single beam sample spectrum was ratioed against the corresponding single beam reference spectrum. Absorbance spectra were generated as the negative logarithm of the ratioed spectra. A total of 197 absorbance spectra were obtained with the 66 sample solutions. All the computer software used for spectral processing was obtained from Professor Gary Small in the Center for Intelligent Instrumentation in the Department of Chemistry at Ohio University (Athens, OH). Subroutines for Fourier filtering and PLS regression were obtained from the IMSL software package (IMSL, Inc., Houston, TX).

Results and Discussion

Absorption Bands. Success of remote NIR spectroscopic sensing depends on the identification and isolation of unique spectral bands for the targeted analytes. These analyte bands must be distinguishable from those of water and other analytes in the sample. Water has large absorption bands with peak maxima at 6876, 5267, and 3800 cm^{-1} in the NIR (12). The spectral region of 5000-4000 cm^{-1} corresponds to an optical window between two large water absorption bands with a minimum around 4500 cm^{-1}.

After taking into account the buffer background, the spectral characteristics of asparagine and glutamine can be compared. Asparagine and glutamine, like other amino acids, have discernible absorption bands around 4390, 4570, and 4700 cm^{-1}. The first band corresponds to the combination bands of aliphatic C-H. The second and third bands correspond to the combination bands of amine N-H. The exact position and width of each band varies for the different amino acids depending on the chemical composition and structure of the side chains.

The chemical differences between glutamine and asparagine result in rather subtle differences in their NIR spectra (see Figure 1). The largest difference is the position and width of the low frequency absorbance bands. This band is centered at 4393 cm^{-1} for glutamine and 4369 cm^{-1} for asparagine. In addition, the asparagine band is wider than that of glutamine. These differences are related to the additional methylene group of glutamine which corroborates earlier findings that NIR bands around 4400 cm^{-1} are associated with the first combination bands of C-H vibrational transitions (8, 11, 15). All the other spectral features are similar in appearance with only minor differences in position, magnitude and width.

PLS Regression. The whole data set representing 197 spectra from 66 samples was divided into two data sets with one for calibration and one for prediction. The calibration data set was composed of all spectra corresponding to 50 samples which were selected randomly from the whole data set. A total of 149 spectra were in the calibration set. The remaining spectra (48 spectra from 16 samples) were placed in the prediction set.

As is reported (10), when applying the PLS algorithm to correlate spectral variations with concentration variations, there should not exist any correlation between the concentrations of the two analytes. The combinations used in this study were built by random selection. Figure 2 presents the correlation plot between the concentrations of asparagine and glutamine in the sample solutions. This plot clearly shows there is no correlation between these two compounds. The r-square value of a linear regression analysis of the entire data set is 0.001. The corresponding r-square values are 0.034 and 0.266 for the calibration and prediction data sets, respectively.

For PLS regression, the spectral range and number of PLS factors are the two crucial parameters. Five spectral ranges were tested for both asparagine and glutamine. They are 4800-4250, 4700-4320, 4650-4320, 4450-4320, and 4700-4450 cm^{-1}. The first three ranges incorporate both the 4570 and 4390 cm^{-1} absorption bands, but differ in the width of the spectral range. The fourth and fifth ranges focus on the 4390 and 4570 cm^{-1} absorption bands, respectively.

The optimum number of PLS factors were determined for each spectral range and each analyte. Table I presents the results of this calculation. For each spectral range and each analyte, the number of factors listed corresponds to the minimum mean standard error of prediction (SEP) (13, 14). An example of our finding is presented in Figure 3a which shows the standard errors of calibration (SEC) and prediction (SEP) as functions of the number of PLS factors used for glutamine with the spectral range of 4800-4250 cm^{-1}. Initially, both the SEC and SEP drop sharply as more of the spectral variation in the calibration data set is incorporated into the model. Little improvement in either the SEC or SEP is observed after 8 factors. As expected, the SEC continues to drop and the SEP begins to increase as the number of factors is increased further. The slight increase in the SEP indicates that the data is overmodelled with too many factors.

The effect of spectral range is typified by the curves presented in Figure 3b which

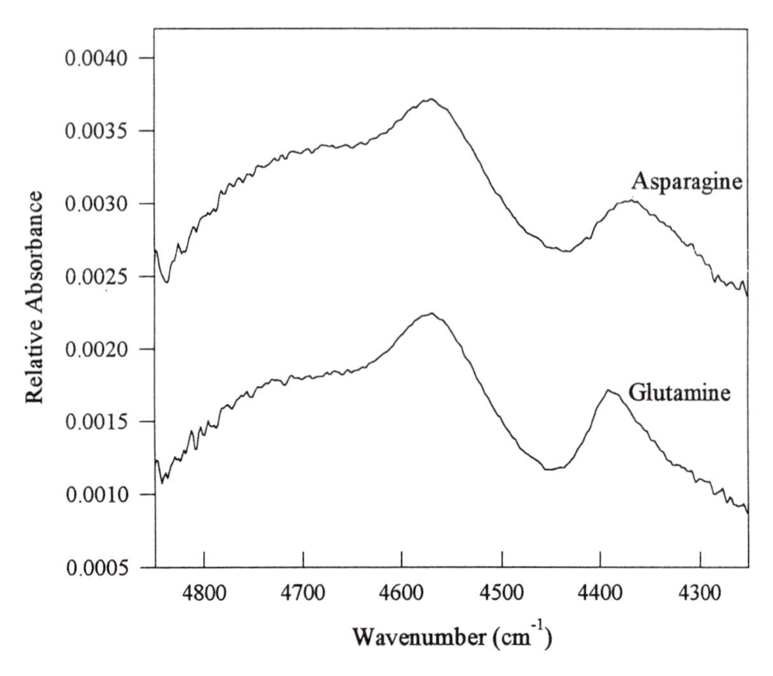

Figure 1. NIR spectra of 10 mM asparagine and 10 mM glutamine in phosphate buffer.

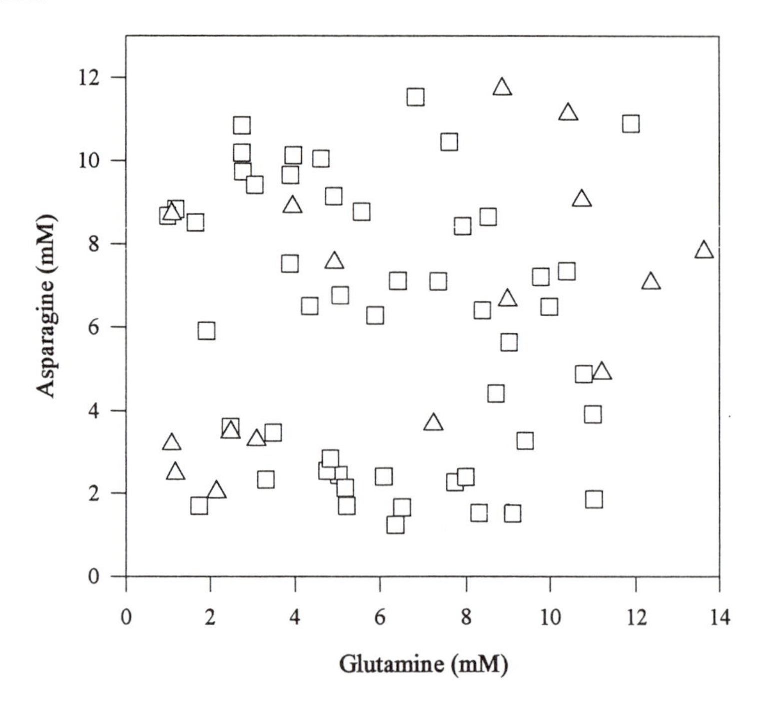

Figure 2. Correlation plot of asparagine and glutamine concentrations in sample solutions for calibration (open square) and prediction (open triangle).

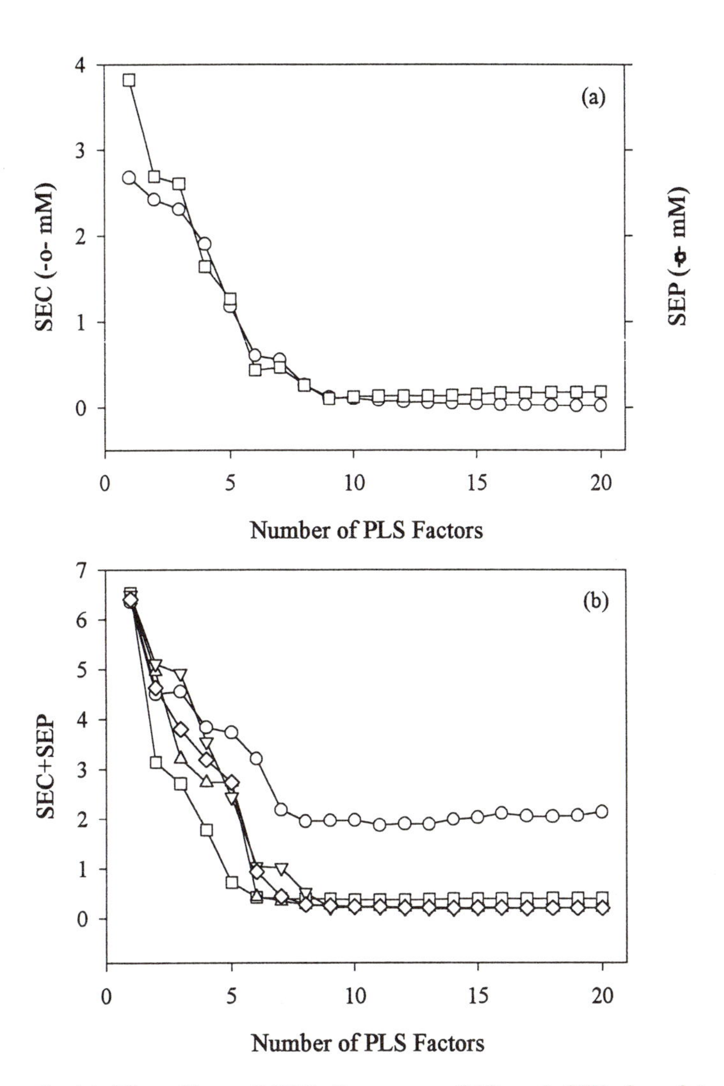

Figure 3. (a) The effect of PLS factors on SEC and SEP for glutamine measurement in the spectral range of 4800-4250 cm^{-1}; (b) The effect of PLS factors on the sum of SEC and SEP of glutamine measurements in the spectral range of 4700-4450 (-○-), 4450-4320 (-□-), 4600-4320 (-△-), 4800-4250 (-▽-) and 4700-4320 (-◇-) cm^{-1}.

show the sum of SEC and SEP as a function of PLS factors for each spectral range. The widest range (4800-4250 cm^{-1}) provides the smallest SEC and SEP and nearly equivalent model performance is obtained with the other ranges that include both spectral bands (4700-4320 and 4650-4320 cm^{-1}). The best spectral range corresponds to the lowest SEP and the fewest number of PLS factors. For glutamine, the best performance corresponds to the 4800-4250 cm^{-1} spectral range and 9 factors. The 4800-4250 cm^{-1} range also gave the lowest SEP for asparagine but 9 factors are needed.

The smaller spectral ranges of 4700-4320 and 4650-4320 cm^{-1} yield slightly larger values for both SEC and SEP. The poorer performance of these narrower spectral ranges is likely caused by less information being provided to the PLS algorithm. Models generated from only one absorption band yield worse results compared to the models built with both absorption bands. Because the major spectral differences are about the 4390 cm^{-1} absorption band, the 4450-4320 cm^{-1} spectral range outperforms the 4700-4450 cm^{-1} range and nearly matches the 4800-4250 cm^{-1} range. It is important to point out that the 4450-4320 cm^{-1} range required the fewest number of PLS factors. More factors are needed to model subtle variations around the 4580 cm^{-1} bands. Similarity in the spectral features around the 4580 cm^{-1} bands is reflected in the relatively large errors obtained when only this band is used in the analysis. Nevertheless, some quantitative information is obtained from the 4580 cm^{-1} region as is evident in the improvement in model performance by using both bands relative to only the 4390 cm^{-1} bands.

The resulting concentration correlation plots are provided for the best glutamine model in Figure 4 where Figures 4a and 4b correspond to the calibration and prediction data sets, respectively. Linear regression analysis of these data yields slopes of 0.998 $\pm$ 0.003 and 1.001 $\pm$ 0.004 as well as y-intercepts of 0.009 $\pm$ 0.116 and -0.009 $\pm$ 0.104 mM for the calibration and prediction data sets, respectively. This calibration model is capable of predicting glutamine concentrations with a 0.10 mM SEP and a 2.19% mean percent error.

For asparagine, the best calibration model corresponds to the spectral range of 4800-4250 cm^{-1} and 12 factors. The corresponding concentration correlations are presented in Figure 5. Figure 5a corresponds to the calibration data set and Figure 5b corresponds to the prediction data set. Linear regression analysis of these data yields slopes of 0.999 $\pm$ 0.003 and 1.039 $\pm$ 0.008 and y-intercepts of 0.01 $\pm$ 0.10 and -0.23 $\pm$ 0.16 mM for the calibration and prediction data sets, respectively. This calibration model indicates a SEP of 0.20 mM and a mean percent error of 3.48%.

PLS Regression with Fourier Filtering. Digital filtering before PLS regression has been used to improve prediction accuracy by enhancing the signal-to-noise ratio (12-15). Fourier filtering is effective in removing high frequency noise and reducing baseline variations associated with raw spectra. During the Fourier filtering process, a raw absorbance spectrum is transformed into a digital frequency domain spectrum by a Fourier transformation. After applying an appropriate Gaussian shaped filter response function, the altered spectrum is converted back to its original data domain. The Gaussian response function is defined by its mean position along the digital frequency axis and its standard deviation, or width of the filter. The mean position must correspond to the molecular absorption features and the standard deviation width must allow a maximum amount of the analyte-dependent information to pass through the filter. An ideal Fourier filtering process can selectively pass the analyte-dependent information while retaining unwanted features, such as baseline variation and noise.

Optimum Fourier filters were identified individually for glutamine and asparagine. The optimum mean position and standard deviation width were obtained by testing various combinations and comparing the quality of the resulting PLS calibration model. Details of this procedure have been provided elsewhere (13-15). For this type of analysis, the data set must be divided into sets for calibration, monitoring and prediction. In this study, 34 samples (102 spectra) were selected randomly from the whole data set for calibration. Of the remaining data, 16 samples (47 spectra) were selected randomly for monitoring and the remaining 16 samples (48 spectra) were used to assess prediction ability.

The ideal filter parameters can be obtained from a 3-dimensional surface map where model performance is plotted as functions of Gaussian mean and standard deviation (14). Figure 6 shows two representative maps, one each for glutamine (Figure 6a) and asparagine (Figure 6b). For glutamine, the spectral range is 4800-4250 cm^{-1} and the number of PLS factors is 9. For asparagine, 12 PLS factors and a spectral range of 4800-4250 cm^{-1} are used. In both cases, the mean position ranged from 0 to 0.10f with a step size of 0.002f and the standard deviation varied from 0 to 0.02f with a step size of 0.001f. The optimum mean, standard deviation pairs are (0.02f, 0.003f) for glutamine and (0.025f, 0.005f) for asparagine. The optimum filter parameters for each spectral range are summarized in Table II.

The optimum filter parameters for glutamine and asparagine are similar which is reasonable considering the similarity in their NIR spectra. Filters for one analyte will pass the spectral information related to the other. Therefore, it is not possible to differentiate these compounds by filtering alone and multivariate analysis must be used to achieve selectivity. Table II summarizes the calibration model performance achieved by combining the Fourier filtering step with PLS regression.

Comparison of the calibration performance with and without Fourier filtering reveals that this filtering step does not significantly enhance model performance when the wide spectral ranges are used. The biggest benefits of the Fourier filtering step are obtained with the 4450-4320 and 4700-4450 cm^{-1} spectral ranges. The SEP for glutamine decreases from 0.22 to 0.12 mM and from 1.27 to 0.12 mM for these spectral ranges, respectively. Similar improvements are observed for asparagine (0.39 to 0.27 mM and 1.32 to 0.22 mM, respectively). The reduction of noise is more relevant over the narrower spectral ranges which contain less analyte-dependent information.

Calibration Models. The best calibration model for glutamine was obtained by PLS regression with Fourier filtering over the 4650-4320 cm^{-1} spectral range with 8 PLS factors and a Fourier filter with a mean of 0.02f and a standard deviation of 0.003f. This model is characterized by a SEC, SEP, and mean percent error of 0.13 mM, 0.10 mM, and 2.00%, respectively. The concentration correlation plots of this best calibration model are presented in Figure 7. Regression analysis indicates

Table I. Results from calibration models for asparagine and glutamine with PLS regression alone.

Glutamine

Spectral range(cm^{-1})	*Number of PLS factors*	*SEC (mM)*	*SEP (mM)*
4800 - 4250	9	0.12	0.10
4700 - 4320	8	0.15	0.12
4650 - 4320	8	0.15	0.12
4450 - 4320	6	0.21	0.22
4700 - 4450	9	0.70	1.27

Asparagine

Spectral range(cm^{-1})	*Number of PLS factors*	*SEC (mM)*	*SEP (mM)*
4800 - 4250	12	0.11	0.20
4700 - 4320	10	0.13	0.21
4650 - 4320	10	0.14	0.22
4450 - 4320	7	0.33	0.39
4700 - 4450	7	1.11	1.32

Table II. Results from calibration models for asparagine and glutamine based on PLS regression combined with Fourier filtering.

Glutamine

Spectral range (cm^{-1})	*Mean*	*Standard deviation*	*Number of PLS factors*	*SEC (mM)*	*SEP (mM)*
4800 - 4250	0.02	0.003	8	0.13	0.10
4700 - 4320	0.02	0.002	12	0.13	0.11
4650 - 4320	0.02	0.003	8	0.13	0.10
4450 - 4320	0.018	0.002	6	0.15	0.12
4700 - 4450	0.02	0.002	8	0.16	0.12

Asparagine

Spectral range (cm^{-1})	*Mean*	*Standard deviation*	*Number of PLS factors*	*SEC (mM)*	*SEP (mM)*
4800 - 4250	0.02	0.005	10	0.14	0.18
4700 - 4320	0.02	0.003	6	0.17	0.21
4650 - 4320	0.02	0.002	5	0.20	0.28
4450 - 4320	0.018	0.001	9	0.22	0.27
4700 - 4450	0.018	0.002	4	0.19	0.22

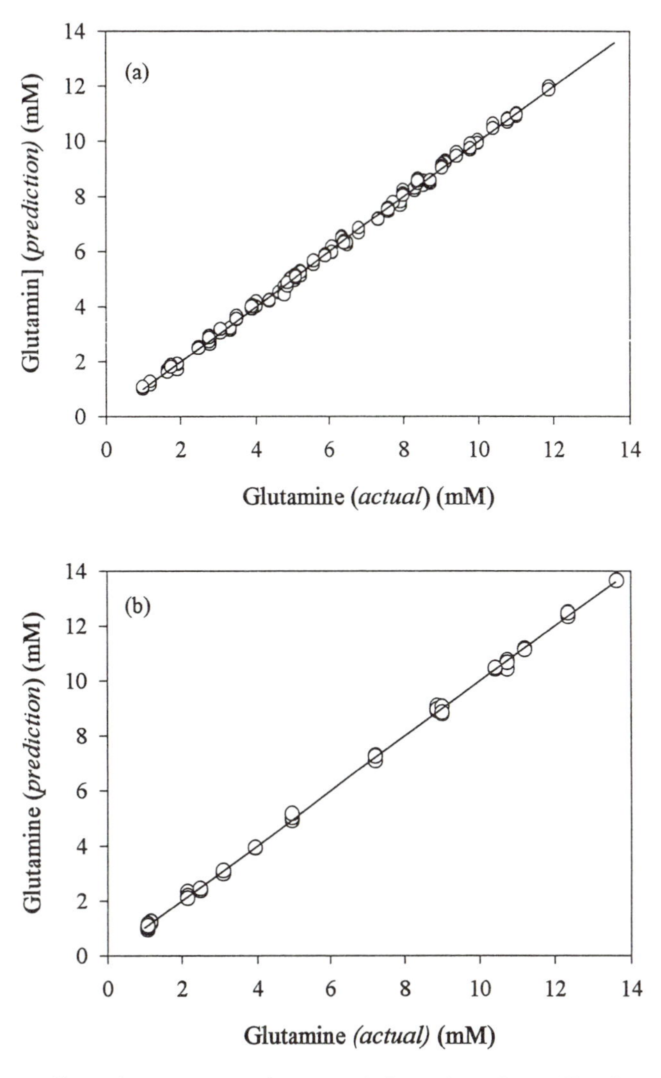

Figure 4. Glutamine concentration correlation plots for calibration (a) and prediction (b) with a spectral range of 4800 to 4250 cm^{-1} and 9 PLS factors. Solid line is the ideal unity correlation.

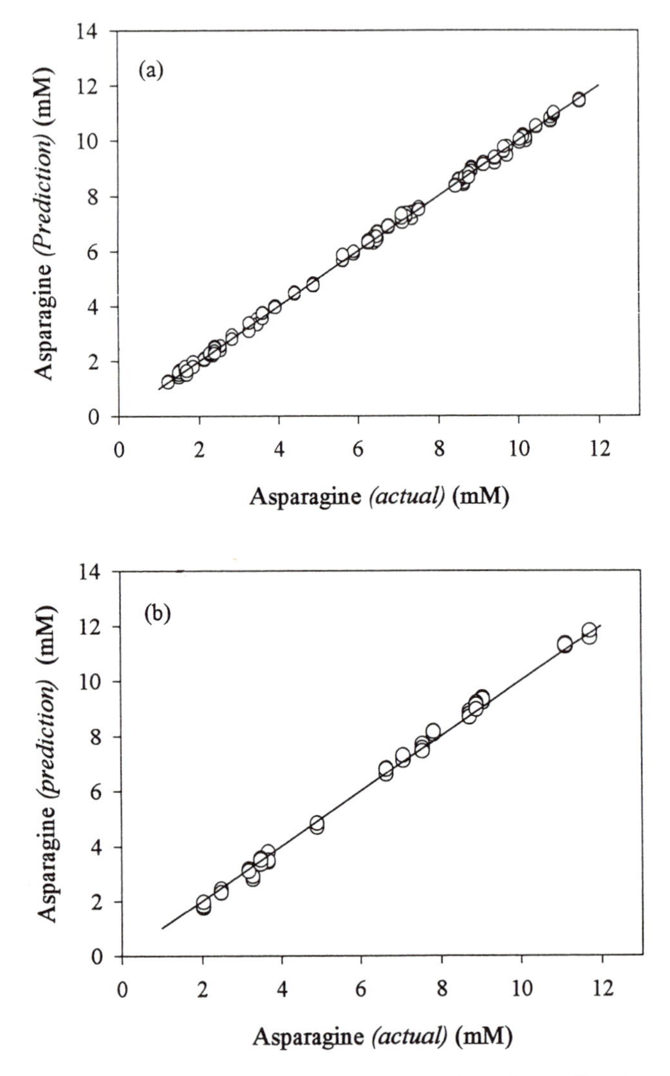

Figure 5. Asparagine concentration correlation plots for calibration (a) and prediction (b) with a spectral range of 4800 to 4250 cm^{-1} and 12 PLS factors. Solid line is the ideal unity correlation.

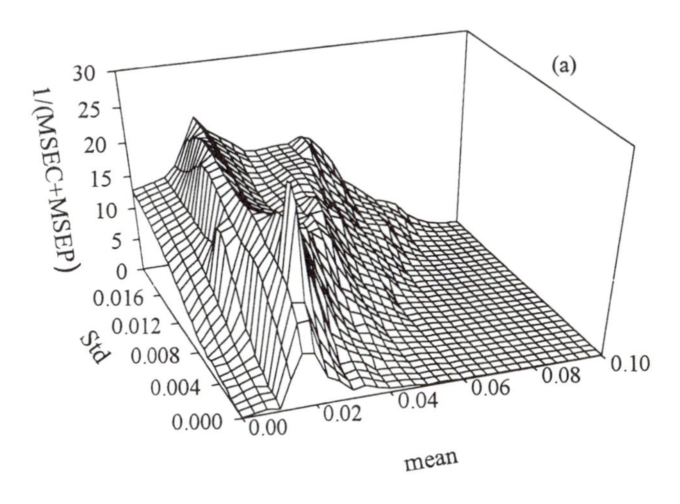

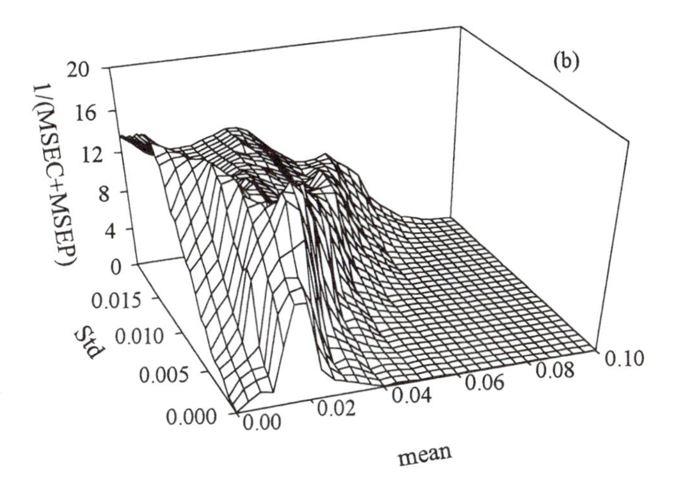

Figure 6. Response surface plots for the optimization of Fourier filter parameters for glutamine (a) and asparagine (b) with a spectral range of 4800-4250 cm^{-1} as well as 9 PLS factors for glutamine and 12 PLS factors for asparagine.

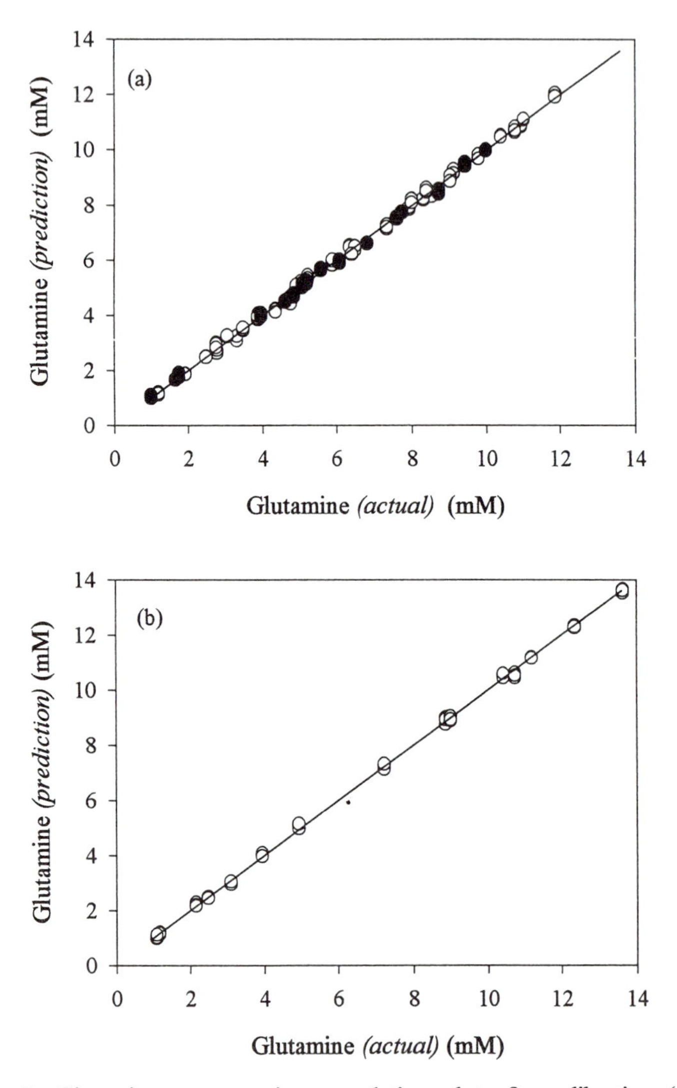

Figure 7. Glutamine concentration correlation plots for calibration (a) and prediction (b) by PLS regression with Fourier filtering in a spectral range of 4650-4320 cm^{-1} with 8 PLS factors. Solid line is the ideal unity correlation. Open circles and solid circles represent the calibration and monitoring data sets, respectively.

slopes of 0.998 ± 0.004 and 0.997 ± 0.003 as well as y-intercepts of 0.01 ± 0.13 and 0.01 ± 0.10 mM for the calibration and prediction data sets, respectively.

The best results for asparagine were also obtained by PLS regression with Fourier filtering over the 4800-4250 cm^{-1} spectral range with 10 factors and a filter with a mean 0.02*f* and a standard deviation 0.003*f*. The corresponding SEC, SEP, and mean percent error values are 0.14 mM, 0.18 mM, and 2.50%, respectively. The concentration correlation plots are shown in Figure 8. Regression analysis yields

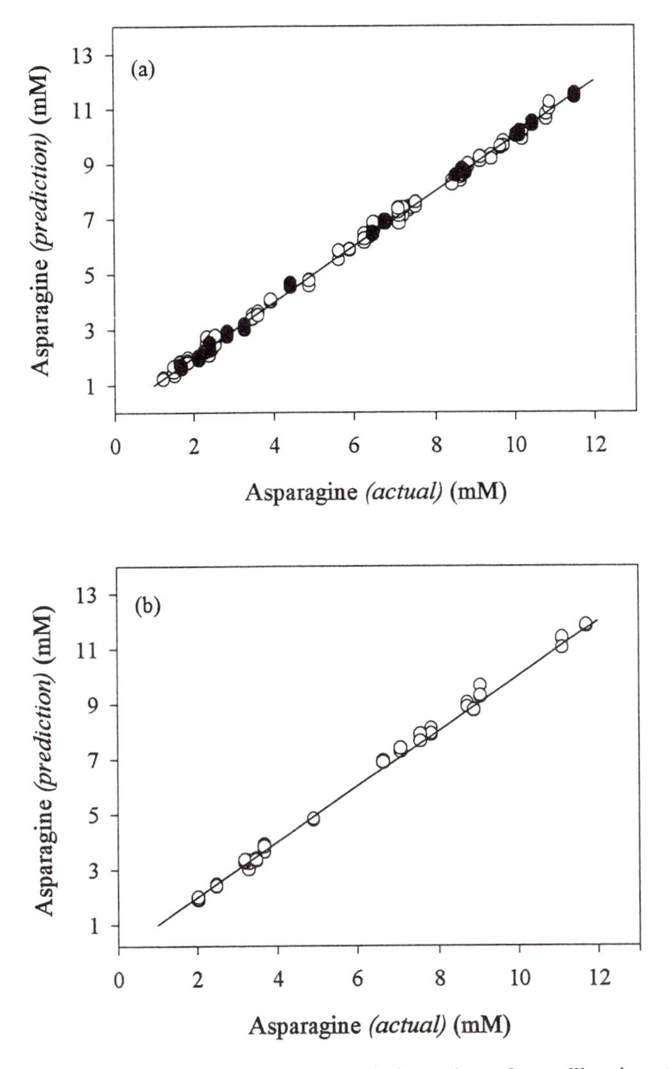

Figure 8. Asparagine concentration correlation plots for calibration (a) and prediction (b) by PLS regression with Fourier filtering in a spectral range of 4800-4250 cm^{-1} with 10 PLS factors. Solid line is the ideal unity correlation. Open circles and solid circles represent the calibration and monitoring data sets, respectively.

slopes of 0.998 ± 0.004 and 1.025 ± 0.008 as well as y-intercepts of 0.01 ± 0.14 and -0.07 ± 0.16 mM for the calibration and prediction data sets, respectively.

The effect of glutamine on the accuracy of predictions for asparagine has been examined by inspecting a residual plot where the residuals for each asparagine prediction measurement are plotted versus the concentration of glutamine present in that sample. These plots are presented in Figure 9 for glutamine and asparagine. In both cases, the residuals do not vary according to the concentration of the companion compound. Residuals for glutamine, for example, are essentially the same regardless of the asparagine concentrations.

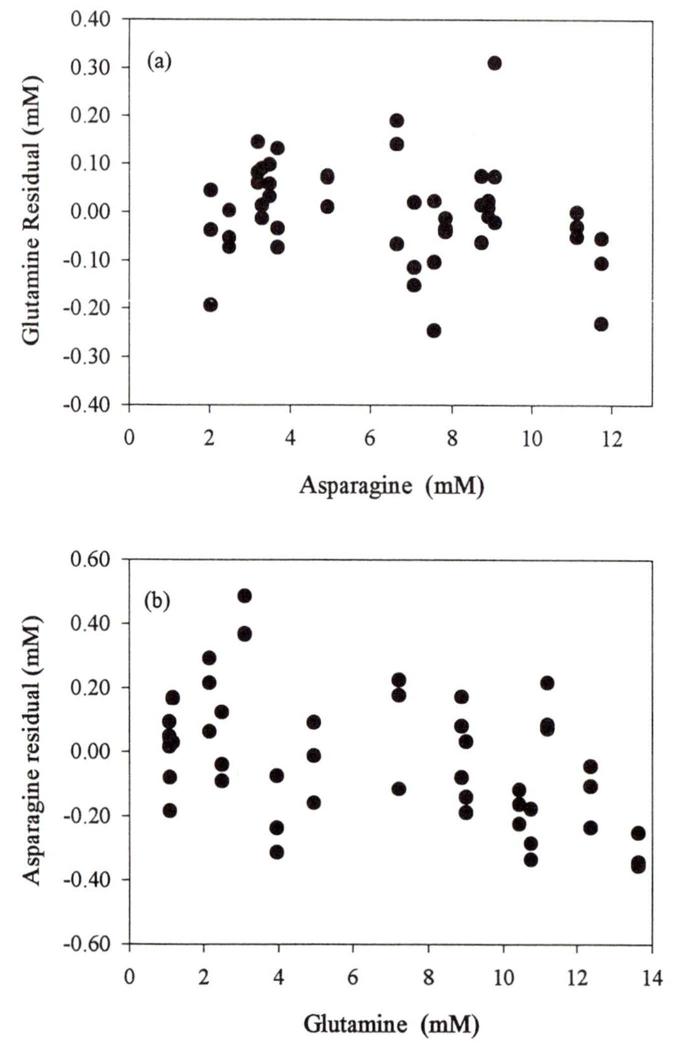

Figure 9. The effect of one analyte on the measurement of the other: (a) Effect of asparagine concentration on glutamine measurements and (b) Effect of glutamine concentration on asparagine measurements.

Conclusion

Asparagine and glutamine can be measured simultaneously by NIR spectroscopy with standard errors of prediction of only 0.18 mM and 0.10 mM as well as mean percent errors of 2.50% and 2.00%, respectively. This level of analysis is possible inspite of the chemical and spectroscopic similarities between these two amino acids. The successful measurement of these compounds suggests that NIR spectroscopy is capable of excellent selectivity which will be critical for further expanding this methodology for measuring multiple components in the complex matrices associated with bioreactors.

Acknowledgment

We thank Professor Gary W. Small in the Center for Intelligent Chemical Instrumentation at Ohio University for supplying the data analysis software and Geng Lu for his assistance in analyzing spectra. The financial support by the National Institutes of Health (NIDDK-45126) is also acknowledged.

Reference

1. Doltzlaw, G., and Weiss, M. D., *Chem. Eng. Prog.* 1993, *89(9)*, 42-45.
2. Kleman,G. L., Chalmers, J. J., Luli, G. W., and Strohl, W. R., *Appl. Environ. Microbiol.* 1991, *57*, 910-917.
3. Dairaku, K., Yamasaki, Y., Kuri, K., Shioya, S., and Takamatsu, T., *Biotechnol. Bioeng.* 1981, *23*, 2069-2081.
4. Glacken, M. W., Fleischaker, R. J., and Sinskey, A. J., *Biotechnol. Bioeng.* 1986, *28*, 1376-1389.
5. Reuveny, S., Kim, Y. J., Kemp, C. W., and Shiloach, J., *Biotechnol. Bioeng.* 1993, *42*, 235-239.
6. Fike, R., Kubiak, J., Price, P., and Jayme, D., *BioPharm.* 1993, *6(8)*, 49-54.
7. Vaccari, G., Dosi, E., Campi, A.L., Gonzalez-Vara y R., A., Matteuzzi, D. and Mantovani, G., *Biotechnol. Bioeng.* 1994, *43*, 913-917.
8. Zhou, X., and Arnold, M. A., unpublished data.
9. He, M., Lorr, D., and Wang, N. S., Near-infrared spectroscopy for online bioreactor monitoring, presented at the 1993 AIChE National Meeting, St. Louis, MO. 1993.
10. Chung, H., Arnold, M. A., Rhiel, M., and Murhammer, D. W., *Applied Biochem. and Biotechnol.* 1995, *50*, 109-125.
11. Stark, E., *Proc. SPIE 8th Internat. Conf. Fourier Transform Spectrosc.* 1991, *1575*, 70-86.
12. Arnold, M. A., and Small, G. W., *Anal. Chem.* 1990, *62*, 1457-1464.
13. Marquardt, L. A., Arnold, M. A., and Small, G. W., *Anal. Chem.* 1993, *65*, 3271-3278.
14. Small, G. W., Arnold, M. A., and Marquardt, L. A., *Anal. Chem.* 1993, *65*, 3279-3289.

15. Hazen, K. H., Arnold, M. A., and Small, G. W., *Applied Spectrosc.* 1994, *48(4)*, 477-483.
16. Krieger, J. H., Borman, S., Baum, R. M., and Freemantle, M., C&EN 1995, March 20, 40-44.

RECEIVED July 20, 1995

Chapter 13

An Expert System for the Supervision of a Multichannel Flow Injection Analysis System

B. Hitzmann, R. Gomersall, J. Brandt, and A. van Putten

Institut für Technische Chemie, Universität Hannover, Callinstrasse 3, 30167 Hannover, Germany

The reliable operation of complex process analysers such as a multi channel FIA system requires permanent supervision. This task can be realised by knowledge-based systems. In this contribution a real-time expert system is presented for the supervision of a multi channel FIA system applied for the on-line bioprocess monitoring of glucose, maltose, polysaccharide, ammonium and alkaline protease measurements. The knowledge-based system combines numerical data analysis with symbolic knowledge processing. Special conditions of real-time knowledge-based systems and the basic structure of the knowledge base are illustrated. Examples of typical faults of the multi channel FIA system are given to explain how on-line symbolic knowledge processing can be combined with numerical analysis of data to perform a fast and reliable fault detection and fault diagnosis.

The complexity of multi channel flow injection analysis systems (FIA systems) implies an increased potential of faults. Thus, an extensive automation of FIA systems, as it is required especially for applications in industrial process control, has to include the automation of supervising functions. However, the fast and reliable detection and diagnosis of faults in FIA systems requires a high degree of knowledge and experience (know-how). This a-priori knowledge has to be transformed into a computer processable form, to take advantage of it for the automation of FIA systems. Transferring it into the form of a knowledge-based system and combining it with information received on-line from the recorded detector signal enables automatic operational supervision of FIA systems.

Various programs have been developed for the automation of flow injection analysis (FIA) systems and many of these are already employed for industrial applications [1]. None of these automation systems consider the complexity of FIA systems with its increased potential of faults. Until now only few investigations have been published, concerning fault detection in flow injection analysis. Giné et al. [2] investigated possible faults of a certain FIA system, the influence of the faults on the

0097–6156/95/0613–0133$12.00/0

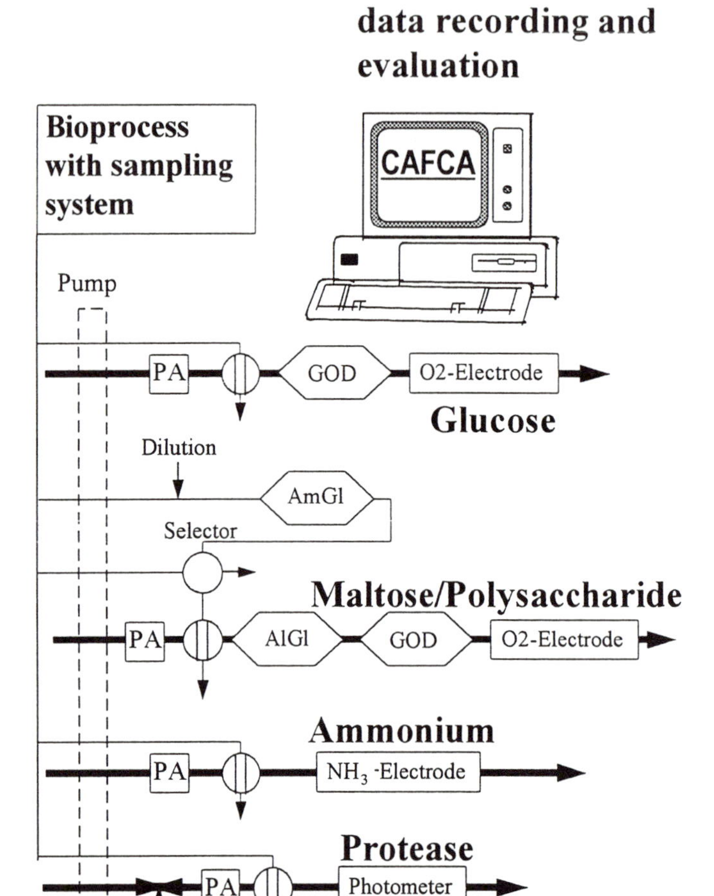

GOD glucose oxydase
AmGl amylo glucosidase
AlGl α–glucosidase
PA puls absorber

Figure 1 Schematic of the multi channel FIA system

measurement signal as well as hints for the remedy for the faults. They emphasise that for slow reaction an evaluation based on the peak height is more precise than an evaluation based on peak area. The error analysis in FIA systems was systematically investigated by Chen and Zeng [3]. They proposed a technique for the detection of faults caused by air bubbles in the flow system. This technique based on the analysis of the first derivative. Szostek and Trojanowicz [4,5] used a digital filter as well as the Fourier transformation of a multiple injection signal [6] to remove high-frequency noise.

Most of these systems used numeric techniques for the fault detection. However, especially for on-line applications the reliable operation of FIA systems demands a high degree of knowledge and experience. The knowledge of the FIA operator enables him to relate specific symptoms of faults in the measurement signal to their cause. To process this kind of knowledge within an automation system, on-line knowledge-based systems are used [7, 8]. With knowledge-based systems the heuristics of experienced FIA operators can be applied to detect and redress faults [9]. This becomes even more important the more complex the FIA system is. Therefore multi channel FIA systems demand the application of knowledge based systems.

This contribution presents a real-time knowledge-based system for the supervision of a multi channel FIA system applied in the on-line bioprocess monitoring of glucose, maltose, polysaccharide, ammonium, and alkaline protease measurements. In this contribution the special conditions of fault detection in a multi channel FIA system are explained. Examples of typical faults are given to explain how on-line symbolic knowledge processing can be combined with numerical analysis of data to perform a fast and reliable fault detection and fault diagnosis.

Materials and Methods

Figure 1 shows the structure of the multi channel FIA system for the simultaneous determination of five process parameters during the cultivation of *Bacillus licheniformis*. The measured components are glucose, maltose, polysaccharides, ammonium and protease. In the first FIA-channel glucose is determined using a oxygen electrode which registers the oxygen utilisation caused by the enzymatic conversion of glucose with the enzyme glucose oxidase. In the second FIA channel alternating either maltose or polysaccharides are converted to glucose using α-glucosidase or amyloglucosidase. The sum of the produced glucose and the glucose which is already in the sample is detected as described for channel one. From the difference between channel two and channel one either the maltose or the polysaccharide concentration is determined. Using the third channel ammonium is determined. An alkaline carrier stream solution causes a pH shift within the injected sample, so that ammonia is produced which is detected by an ammonia electrode and can be correlated to the ammonium concentration. With the fourth channel the activity of the protease subtilisine Carlsberg is determined using the stopped-flow FIA technique. The protease is used to convert N-CBZ-valin-p-nitrophenylester to p-nitrophenole, whose absorption is measured photometrically at 340 nm and can be correlated to the protease activity.

The control of the multi channel FIA system as well as the data recording (1 Hz), evaluation, visualisation, and storage are performed by the automation system

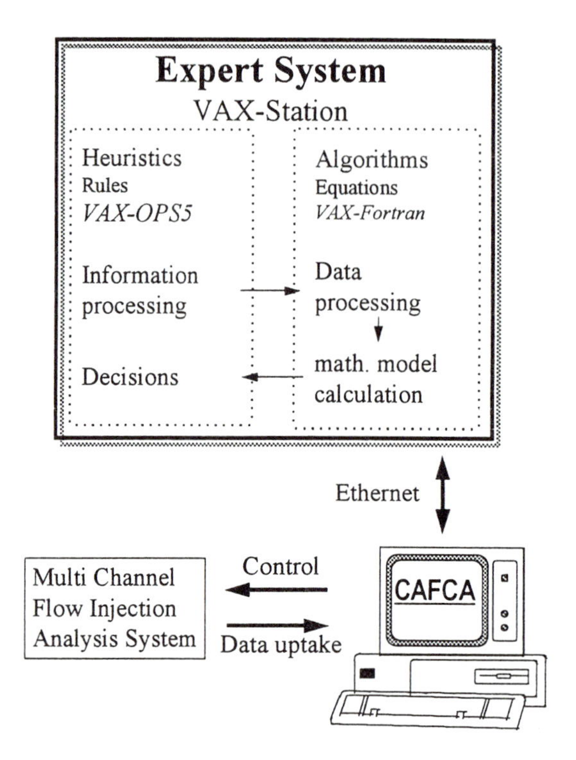

Figure 2 Structure of the expert system

Table I Typical faults observed in the FIA system

Faults of the				
sampling system	**flow system**	**reaction system**	**detector system**	**automation system**
•plugging of filtration device	•pulsating flow rate	•loss of enzyme activity	•loss of main voltage supply	•hardware breakdown
•malfunction of calibration valve	•jammed tubes	•fluctuating temperature	•ageing of membranes	•loss of main power supply
•microbial contamination of the sampling device	•air bubbles	•enzyme inhibitors within sample	•damaging of membranes	•control or data wire disconnected
•wrong standard solutions	•plugging of tubes	•changes in carrier composition	•photometer disturbed by light	•wrong I/O ports selected
•unstable dilution rate	•malfunctioning injection valve	•wrong sensitivity	•fused power supply	•wrong input of concentration of standards for calibration
•insufficient standard solution	•leaking fittings	•insufficient carrier solution	•wrong polarisation voltage	•wrong A/D range selected
	•change of flow rate	•sample contains proteases	•signal leaves detection range	•false programming of control
	•burst of tube fittings		•false earth connection	

CAFCA (Computer Assisted Flow Control & Analysis, ANASYSCON, Hannover, Germany), which runs on a MS-DOS computer. The knowledge-based system was developed on a VAXstation 3100 (Digital Equipment Corp., Maynard, MA) and combines numerical data analysis with symbolic knowledge processing. The knowledge based module is implemented with a development tool for production systems - the VAX-OPS5. All numerical algorithms were developed in VAX-FORTRAN. The structure of the knowledge-based system is shown in Figure 2. As can be seen in this Figure the connection between the knowledge-based system and the FIA system is achieved by a fast link (Ethernet, DECnet) via CAFCA.

Results

To develop the knowledge-based system all the knowledge concerning faults of the multi channel FIA systems was acquired. In Table I a list with examples of typical faults is presented, which can be observed with the multi channel FIA system. In order to detect a fault of the FIA system the only source of information which can be exploited by a real-time knowledge-based system is the measurement signal. Therefore, to recognise a disturbance of the signal shape, numerical algorithms are applied to calculate values that characterise the contour of each FIA signal (feature extraction). The features are processed by rules, testing specific limits and interpreting the actual values for a diagnosis of the FIA system state. If a feature deviates from its normal value, a hypothesis is generated by the rules, that a fault might be possible. Therefore, within the expert system numerical procedures and heuristic rules are assembled.

The feature extraction out of the measurement signals takes place in three different cycles. In the first cycle features characterising single data points of the whole FIA signal are calculated. For example, the difference between two data points or the distance of the value of a data point to the maximum amplification range are typical features calculated in this cycle. In the second cycle features characterising a complete FIA signal are calculated. In case of a peak shaped signal the calculated features are, e. g., the baseline, the maximum and the width at 5, 50 and 95% height. In a third cycle the knowledge-based system calculates trends of features. To characterise sudden changes of a feature the value is compared with the value of the previous cycle; for slow changes of a feature the values gradient over several cycles is calculated. Features used by the expert system to recognise faults can be seen in Table II.

The following example illustrates how a fault can be detected by the analysis of trends of these features. In Figure 3 the measurement signals of some injection cycles are shown which were detected by the oxygen electrode of the glucose channel. The first and second peak are registered without any trouble. However, after the peak with the cycle number 26 was recorded, a breakdown of the supply of carrier stream solution occurred. Because of this fault the peristaltic carrier stream pump pumped air into the tubing system instead of carrier solution. First of all this fault causes a reduced carrier stream and after about 10 minutes of pumping air all the liquid has left the tubing system. This will cause a rapid decrease in the recorded signal as can be seen in Figure 3 shortly after the peak number 30. This fault can be detected by analysis of three characteristic features: the residence time, the half width, and the drift. These features are routinely calculated from the measurements of each

Table II Examples of features used to recognise faults

fault	influence on the FIA system	main feature for the recognition
electronic disturbance	enlarged signal/noise ratio	difference between single data points
air bubbles	spikes	difference between single data points
breakdown of the pump	flow rate stops immediately	difference between single data points, no peak detected
injektor switched false	reduction of peak height	trend of peak height
matrix influence	tailing	trend of the peak widthat 50% of the peak height
plugging of injection valve	enlarged pressure in the flow system, reduction of the sample volume, reduction of flow rate	extrema in baseline shortly after injection
plugging within carrier transport	reduction of flow rate	residence time
concentration in sample unexpectedly high	measurement exceeds linear range	statistical moments
burst tube fitting	no carrier flow	drift of baseline
infection in enzyme cartridge	additional oxygen usage	trend of the peak width at 5% of the peak height
ageing of the O_2 electrode membrane	pressure sensitivity	trend of the base line noise
ageing of the NH_3 electrode membrane	signal drifts towards maximum amplification range	trend of the base line, trend of the peak height
breakdown of the thermostat	temperature not stable, change of convolution	trend of the base line drift

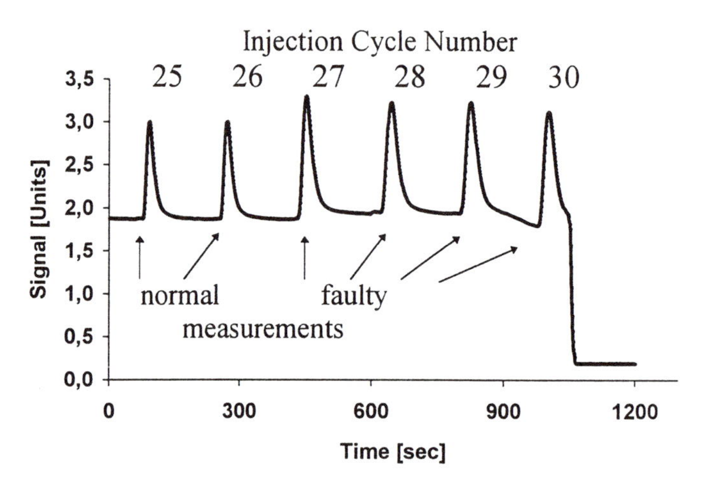

Figure 3 Measurement signals recorded during a breakdown of the carrier stream

individual injection cycle. In Figure 4 the evolution of these features are shown. As can be seen in the Figure all features change their values significantly for the signal of the injection number 27. During the feature extraction and analysis of this cycle the expert system recognises first the increase of the residence time. Then the symptom *residence time is increased* is generated. Soon after that rules generate two additional symptoms: *half width increased* and *positive drift*. From these three symptoms the expert system concludes that a fault has occurred in the flow system, which has reduced the pumping rate. This can have many causes. However, from the rapid increase of the features values the expert system can conclude, that causes can be rejected which will have a more continual change of the features (such as attrition or soiling). Therefore, the expert system concludes that a breakdown of the supply of carrier solution is most probable. If this fault is recognised by the operator the dead loss of this FIA channel can be prevented.

The example shows, that a fault can be detected based on a fast detection of deviations of the characteristic features as well as the analysis of their values with heuristic rules. As soon as the value of a feature is not within its limits rules will produce a hypothesis that a certain disturbance has been detected (symptom recognition). The presence of a hypothesis activates additional rules to diagnose the cause of the disturbance and to verify or falsify the hypothesis. The recognition of symptoms caused by faults is the ultimate goal of the knowledge-based system, so it can alarm the operator immediately. However, the analysis of the symptoms and the identification of the specific faults is also of great importance. A fundamental problem is to distinguish between faults that reveal themselves in the same or in similar symptoms.

In a single channel supervisory system only one signal serves as the source of information on faults in the FIA system. To be able to differentiate between two faults with similar symptoms and achieve a precise diagnosis the analysis of the trends of features has been found to be the best source of information. The knowledge-based system supervising a multi channel system has access to many signals as source of information. By comparing the state of every FIA channel with each other, disturbances in the system can more precisely be traced back to their cause as well as their location than in a single channel system. The individual FIA channels are not fully independent from each other as they share several components. For example, there is only one on-line sampling device responsible for the sufficient supply of all FIA systems. Faults occurring to this component that is vital for all channels will result in a disturbance of all signals. Furthermore, the measured process variables are from one bioprocess so their values are in direct relation with each other. This is used as additional source of information. For the extension of the single channel knowledge based system a complete new set of rules was required to make use of all this supplementary information.

To illustrate the inference capability of the knowledge based system for a multi channel FIA system an other example is presented. In this example special rules achieve an accurate diagnosis by taken into consideration the a priori knowledge (such as knowledge about the specific structure of the multi channel FIA system) as well as heuristics of the experienced operators. In the example, the fault of a decreased flow rate is also discussed and how it effects different FIA channels. The disturbance of the signal shape is dependant on the specific measuring principal. In

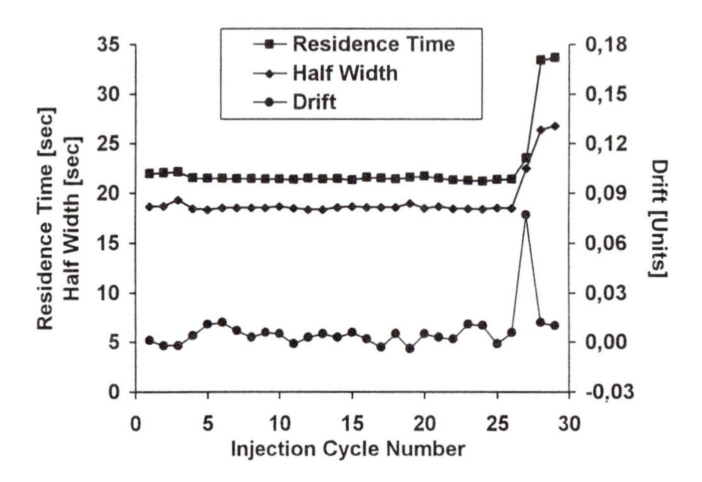

Figure 4 The evolution of three features during an occurrence of the fault

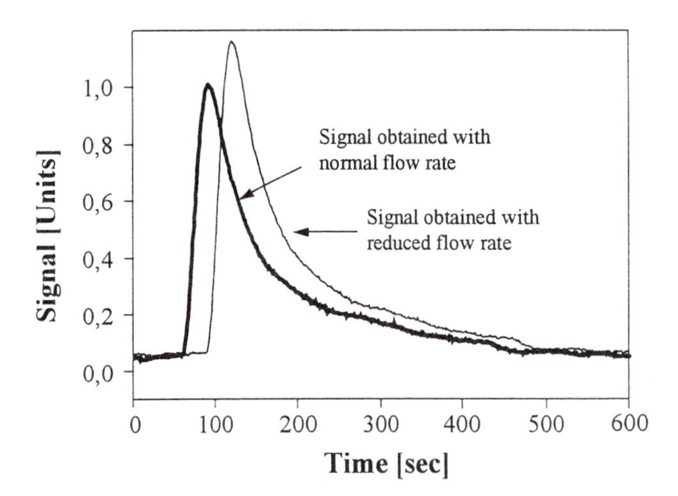

Figure 5 Undisturbed and disturbed signal of a glucose FIA channel

Figure 5 the normal and disturbed signals are shown, that were detected with the glucose channel.

As explained in the first example a reduction of flow rate causes an increase of the residence time of the sampling zone in the manifold and this leads to a higher conversion during the FIA cycle. As a result the glucose concentration is determined to be much higher than it really was. As a further hint that a disturbance has occurred the fact is used, that for this specific bioprocess the glucose concentration should not increase, because it is consumed rather than produced.

In Figure 6 the normal and disturbed signals are illustrated, that were detected with the stopped flow FIA channel for the monitoring of protease. The signal has got a totally different shape than the signal obtained from the channel described above.

In the stopped flow system the carrier flow is stopped by the automation system CAFCA in a timely manner, so that the complete sampling zone is transferred into the reaction chamber. At the end of a FIA cycle the carrier flow is restarted and the sampling zone is washed out of the system by the carrier. The reduction of flow rate in this system causes the sampling zone not to be fully transferred into the detection volume of the photometer. The reaction takes place in a part of the manifold that is not detected by the photometer. As soon as the sampling zone is washed out it passes the photometer which results in a steep positive gradient of the signal. The two rules that detect this characteristic change of the signal shape can be described as follows: *IF* the signal of channel X is a stopped flow FIA signal, *AND* the gradient after the carrier flow is restarted is positive *THEN* create the hypothesis "residence time might have changed for channel X" *AND* calculate the maximum gradient. *IF* a hypothesis "residence time might have changed for channel X" exists *AND* the value of the concentration is smaller than predicted *THEN* create the diagnosis "flow rate reduced for channel X".

For both channels the rules deduce from the abnormality of the separate signals that the flow rate is reduced. A more precise diagnosis is not possible as there are several faults that cause a reduced flow rate. The most probable causes are a reduced pump rate or a jammed carrier tube. To be able to differentiate between these two causes it can be taken into consideration that both channels are supplied by a pump where as each separate channel has its own carrier tube. If the pump rate is reduced then a diagnosis will be produced for both channels. In case of a jammed carrier tube only one signal will be disturbed and therefore the faults can be differentiated. The rule that will diagnose a jammed carrier tube of channel 5 can be described as follows: *IF* a diagnosis "flow rate reduced for channel **5**" exists, *AND* there exists no other diagnosis "flow rate reduced channel X", *THEN* create alarm "Carrier tube channel **5** must be jammed! Most probably pump rate reduced!". In this manner the knowledge-based system detects and analysis deviations from the normal state of the FIA system. Typical faults which can be recognised and identified by the knowledge-based system are as follows:

> air bubbles in the flow system, electrical disturbances such as spikes, reduced carrier flow caused by a jammed carrier tube, reduced carrier flow caused by reduced pump rate, interrupted carrier flow (disconnected fittings), breakdown of the pump, breakdown of the thermostat, signal outside of the amplification range, calibration standards swapped, incorrectly switched injector valve, injector valve plugged by

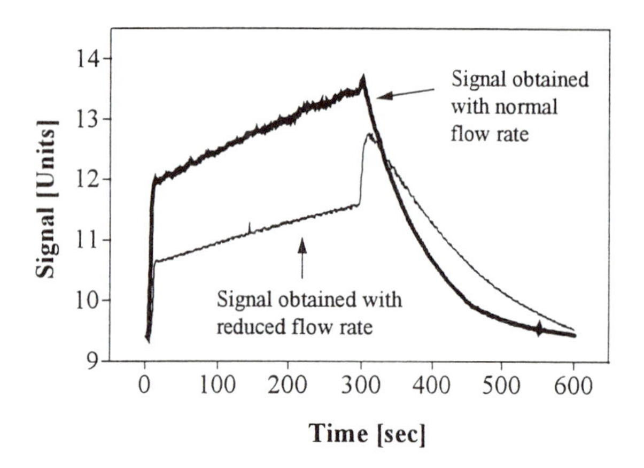

Figure 6 Undisturbed and disturbed signal of the protease stopped flow FIA channel

microorganism growth, damaged electrodes, infection of the enzyme cartridge, influences of the cultivation broth.

However, even if the cause of a fault is not concluded in the right way, the hints of the expert system are a tremendous support for the FIA operator.

Conclusion

Transferring the knowledge and experience of human operators of multi channel FIA systems into the form of a knowledge-based system and combining it with information received on-line from the recorded detector signal enables automatic operational supervision of FIA systems. With the help of this system the operator must not always follow the complex analysis system closely. The knowledge based system provides the operator with information about the state of the analysis system and an immediate report of possible faults is given as soon as a symptom is recognised. Comparing the fault analysis of a multi channel FIA system to a single channel system the precision of diagnosis is much higher because of the enlargement of information. The bottleneck of knowledge-based system development is still the knowledge acquisition. Based on the knowledge of an experienced operator the system is developed. Therefore, the knowledge based system is as skilful as the human expert. However, the knowledge based system will always be ready and attentive, never tired. The benefits of such a system for FIA supervision are lower expenses and higher safety.

References

[1] J. Ruzicka, E.H. Hansen, Flow Injection Analysis, Wiley& Sons, New York, 1988

[2] M.F. Giné, R.L. Tuon, F.J. Krug, M.A.Z. Arruda, Anal. Chim. Acta 261(1992)533

[3] D. Chen, Y. Zeng, Anal. Chim. Acta 235(1990)337

[4] B. Szostek, M. Trojanowicz, Anal. Chim. Acta 261(1992)509

[5] B. Szostek, M. Trojanowicz, Anal. Chim. Acta 261(1992)521

[6] B. Szostek, M. Trojanowicz, Chemometrics Intell. Lab. Syst. 22, 221-228
[7] T. J. Laffey, P. A. Cox, J. L. Schmidt, S. M. Kao, J. Y. Read, AI Magazine 9 (1988)27
[8] D. A. Rowan, AI Expert (August 1989)30
[9] J. Brandt, B. Hitzmann, Anal. Chim. Acta 291 (1994)29

RECEIVED August 11, 1995

Chapter 14

Hybrid Process Modeling for Advanced Process State Estimation, Prediction, and Control Exemplified in a Production-Scale Mammalian Cell Culture

M. Dors, R. Simutis, and A. Lübbert

Institut für Technische Chemie, Universität Hannover, Callinstrasse 3, 30167 Hannover, Germany

Indirect measurements, process state predictions and optimization can be largely improved by exploiting all available knowledge and data about a given production process. This concept has been implemented in an industrial process producing a recombinant protein in mammalian cell culture. The model has been used for indirect measurements and optimization of feeding profiles and was directly used for process control.

As is well-known in practice, the protein folding properties and, hence the product quality is dependent on the cultivation conditions of the host organism. Thus, in order to guarantee products within narrow specification limits, the production process must be kept under tight control (4). An essential prerequisite is accurate monitoring of the process' state and an optimization of the trajectories of the key process variables. Advanced control strategies require to predict the process behaviour at least over time horizons which are needed to influence the process so that the state variables will not escape from the acceptable intervals. Prediction, however, means that the process has to be modelled.

In this article we describe a model used to supervise a process in which a recombinant protein is produced with mammalian cell culture. This model has been used to optimize the process trajectories for the production process.

Structure of the model

Production processes using genetically modified cell systems are not so thoroughly understood that they could be modelled comprehensively with physically based mathematical models only. On the other hand there is much heuristical knowledge available from many productions run in industrial practice. This can be described by correlations or at least by simple rules-of-thumb. The main idea behind the model to be described is to exploit all the knowledge and information which is available, may it be provided by means of mathematical models, heuristic knowledge or even by simple

0097–6156/95/0613–0144$12.00/0

data sets from successful production runs only. The main problem is how to represent and process that knowledge and how to effectively combine the information from different sources so that the different pieces of a priori knowledge can be processed simultaneously.

Shortly, the proposed approach fits the techniques used to monitor, predict and optimize a particular production process to the knowledge available. What is known fuzzily is treated accordingly, while what is known accurately will be treated by means of conventional mathematical models.

The basic idea is first of all, to provide a frame for the model, making use of well-established knowledge and then to fill the gaps with mathematical, heuristical and data driven approaches, depending on the mode the a priori knowledge is available.

The frame we use is the overall mass balance, which is usually represented by a set of ordinary differential equations for the main state variables. An example of a balance equation system for the fed batch culture of mammalian cells discussed here is:

$$dX/dt = r_X - \frac{F}{V} X$$

Biomass, X, production
- Dilution effect through feed

$$dS/dt = - r_S + \frac{F_f}{V}(S_f - S) + \frac{F_z}{V}(S_z - S) - \frac{F_l}{V} S$$

Glucose, S, consumption of the cells
+ Dilution effect of fresh medium
+ Addition through glucose-glutamin feed
- Dilution effect through base addition

$$dG/dt = - r_G + \frac{F_f}{V}(G_f - G) + \frac{F_z}{V}(G_z - G) - \frac{F_l}{V} G - K_d G$$

Glutamine, G, consumption of cells
+ Dilution effect by feed F_f with fresh medium
+ Dilution effect by feed F_z glucose-glutamine feed
- Dilution effect through base addition F_l
- Glutamine disintegration

$$dL/dt = r_L - \frac{F}{V} L$$

Lactate, L, production
- Dilution effect
(In this example no consumption of lactate observed)

$$dA/dt = r_A - \frac{F}{V} A + 0.123^{1)} K_d G$$

Ammonia, A, production
- Dilution effect
- Ammonia development by glutamine disintegration

[1)] experimentally determined for this particular case

$$dP/dt = r_P - \frac{F}{V} P$$

Product, P, formation
- Dilution effect

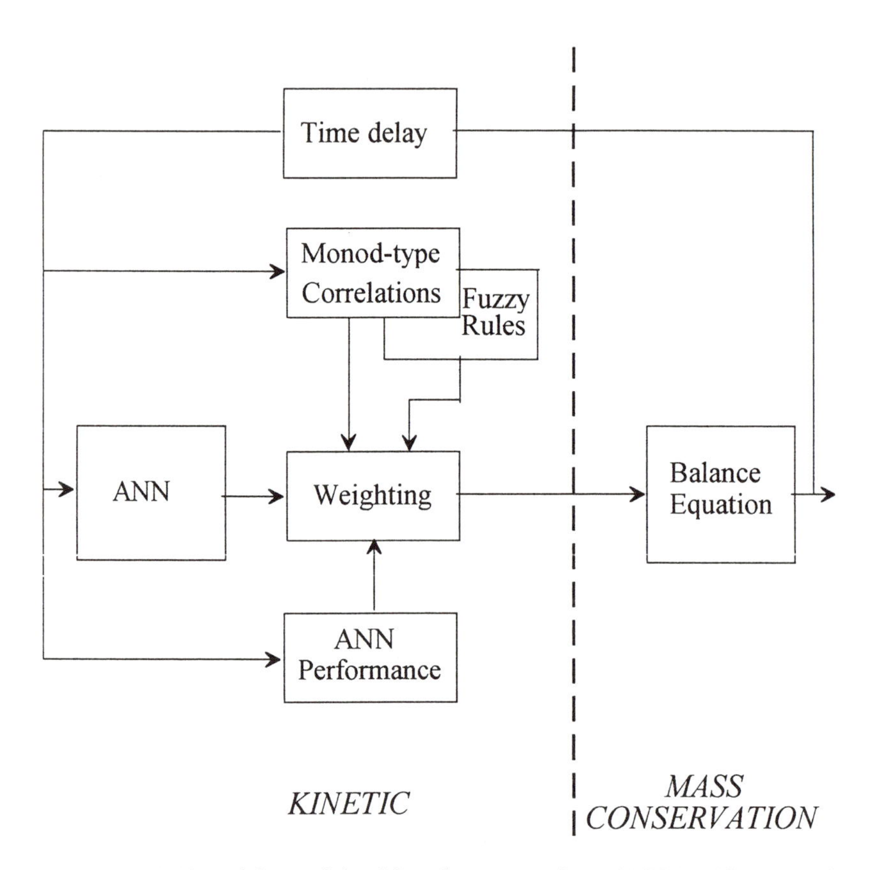

Figure 1. Hybrid model containing Monod-type equations, ANN (Artifical Neural Network), mass balance equations and evidence monitor for the ANN.

$dV/dt = F - F_w$
Volume V change by harvesting

$F = F_f + F_z + F_l \geq 0$
Total feed, F, into the vessel =
Feed, F_f, of fresh medium
+ Feed, F_z, of glucose-glutamine
+ Feed, F_l, of base addition.

The most challenging terms in such a balance equation system are the specific rate μ for growth and the rates r_i for production and/or consumption of the i-th component. These are issues of current discussion and cannot be provided on the same level of accuracy as the balance equations themselves.
Hence we use two sources of knowledge simultaneously:
(i) The textbook approach of mammalian cell kinetics in form of modified Monod-type correlations and
(ii) artificial neural networks trained on the data from the on-going production process.
Figure 1 presents a scheme of the model architecture.

The Monod-type approach used considers two limitation and two inhibition terms:

$\mu = \mu max$ Maximum specific growth rate

$* \frac{S}{S + K_S}$ Limitation of glucose conc.

$* \frac{G}{G + K_G}$ Limitation of glutamine conc.

$* \frac{K_A}{A + K_A}$ Inhibition by ammonia

$* \frac{K_L}{L + K_L}$ Inhibition by lactate

$$r_X = m\ X$$
$$r_S = \left(\frac{r_X}{Y_{X/S}} + m_S \frac{S}{S+K_{mS}}\right) X$$
$$r_G = \left(\frac{r_X}{Y_{X/G}} + m_G \frac{G}{G+K_{mG}}\right) X$$
$$r_L = Y_{L/S}\ r_S$$
$$r_A = Y_{A/G}\ r_G$$
$$r_P = Y_{P/X}\ r_X + b$$

$Y_{X/S}$ = Yield Biomass/Glucose
$Y_{X/G}$ = Yield Biomass/Glutamine
$Y_{L/S}$ = Yield Lactate/Glucose
$Y_{A/G}$ = Yield Ammonium/Glutamine
$Y_{P/X}$ = Yield Product/Biomass
m_S = Maintenance term Glucose/Biomass
m_G = Maintenance term Glutamine/Biomass
The parameters were estimated from available process data.

Both approaches, the classical Monod approach and the neural network approach are limited by their specific constraints. The Monod correlations are restricted to the general form of the equations and the values of the essential parameters and functions (in particular the yield expressions which interrelate the different rates) within them. The artificial neural networks can only be used in situations which have been experienced practically and from which the essential data have been used for the network's training. Both sets of constraints cannot be formulated in a crisp way, they can only be represented in a more or less heuristical way. Because of this fuzzy situation, the constraints are provided by means of rules, formulated with fuzzy variables and processed using fuzzy logics.
It should be stressed at this point that both approaches must be used simultaneously (2,3). Both contain valuable information, however, their relative importance changes due to the particular situation, i.e., from point to point in the state space. In order to adapt the weights of both accordingly, the evidence of the network component has been monitored continuously. In order to determine the areas of experience a cluster analysis has been performed. With a relative importance I_n of the network, the importance of the Monod correlation is $(1-I_n)$. It is straightforward to couple I_n to the amount of data available to train the network in the area immediately around the actual state space position. This can be performed with the evidence measure (1), (Simutis,R., Dors, M., Lübbert, A., Increasing the efficiency of hybrid models in bioprocess supervision and control, in preparation). The simultaneous utilization of mathematical models, heuristic rule systems, and artificial neural networks has been referred to as a hybrid model.
This way of combining a priori knowledge with data from the actual process has the advantage that the hybrid model can be used for state estimation even if only a few measurement data are available. Typically, the data from 3 - 5 production runs are sufficient. Then, obviously, the weight of the neural network component is very low and the software relies on the Monod-type correlation and on the fuzzy rules mainly. With more data the weights of the networks will increase later on.
There are several advantages in using this hybrid technique.
1. Since more knowledge and data is exploited, the accuracy of the model is superior compared to conventional approaches.
2. The model can be implemented with the actually available knowledge and it will learn with its applications, i.e. from additional data when they become available. This can be regarded as an improvement in comparison with pure neural network approaches.
When it comes to industrial applications of that concept, in particular under real-time conditions, it is essential to adapt the generalized, hybrid models to the data which are actually available. Hence, different models are to be used for different applications, e.g., for state estimation and control on one, and state prediction on the other hand. In state estimation one tries to use as many input parameters as are available from the measurements that are related to the output parameters. The main objective is to improve the estimation accuracy. In state prediction, we do not have measurement values for future time points. In this case it proved to be of advantage to use a minimal set of process variables that describe the state of the process in order to minimize the distortions in state predictions. This minimal set is the set of the state variables which are most often different from the directly measured quantities.

Consequently a model interrelating these state variables has been used in process simulation and optimization.

Implementation of the model

The model described was implemented in a software package written in Fortran. It was processed by the real time integrating software platform RISP (Dors.M, Havlik, I., Lübbert, A., RISP Realtime Integrating Software Platform, Report, Institut für Technische Chemie, Universität Hannover, Callinstr. 3, D-30167 Hannover) which runs under the operating system VMS on VAXstations from DIGITAL (DIGITAL, VMS Operating System, V5.5, Digital Equipment Corporation, Maynard, USA). RISP is a realtime data management system which can be regarded as an advanced process control software tool, which furthermore provides all tools necessary to process advanced data analysing software modules. It supports them with the required data may they be historical data from a data base or currently measured data from a running process. The data can be preprocessed in any way which may be necessary for an analysis. RISP furthermore provides a comfortable user surface for data monitoring from the attached process as well as from the actually running analysing programs. User surfaces follow actual software standards. The connection to the production plant is usually made by commercial programmable control systems (PCS) e.g. Siemens S5 or by the front end systems provided by bioreactor manufacturers, e.g. B.Braun DCU. The software has been improved during several industrial applications in different biochemical production processes.
External analysing programs can be coupled to RISP in a convenient way. An extended library is provided to make use of all the data recources and data representation features of RISP.
Figure 2 shows a typical operator interface which allows to overview the process state at a glance. All essential process variables including the on-line estimated ones are represented in form of small continuously updated graphs within so-called push buttons, thus imitating chart recorders. They reflect the immediate history of the different variables over short, medium or long time intervals as well as the characterizing numerical values. Once, the operator recognizes a suspicious change in a graph he can enlarge it to a fully scaled plot by a simple mouse-click on that buttom. RISP also provides all necessary documentation features which are necessary in production plants. In this respect it uses the Digital RDB data base.
Off-line data from the laboratory can be typed in via a user surface which can be operated by the plant personnel. Once the data are acknowledged by the system they are used to update the process models from the sampling time to the actual time. Also, the improved state estimation will be used for an update of all the predictions run during the process.

Application

The software has been applied to a recombinant protein production plant. The mammalian cell culture is processed in a repeated fed batch mode. The essential aim of a model supported control is to continuously estimate the actual state of the culture and, on this information, to predict its behaviour during the next period up to an

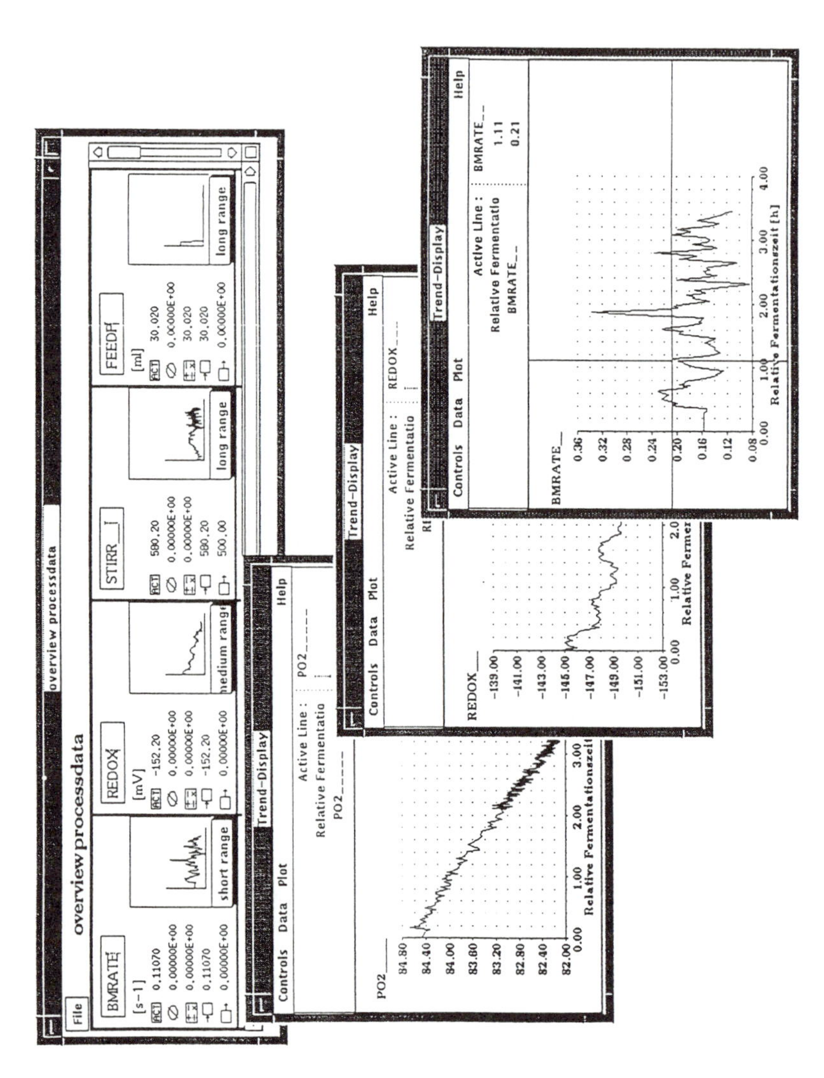

Figure 2. Typical user interface from the RISP software.

optimal harvesting point. The harvesting time and the amount of culture to be replaced by fresh medium must be chosen carefully since failures would result in a significant productivity loss. Moreover the growth in the individual cycles can be significantly influenced by an optimal feeding strategy.
As an example, some typical on-line estimated signal curves are shown in Figure 3a and Figure 3b together with the actual process data which became known later on. As can be seen, the estimations are quite accurate.

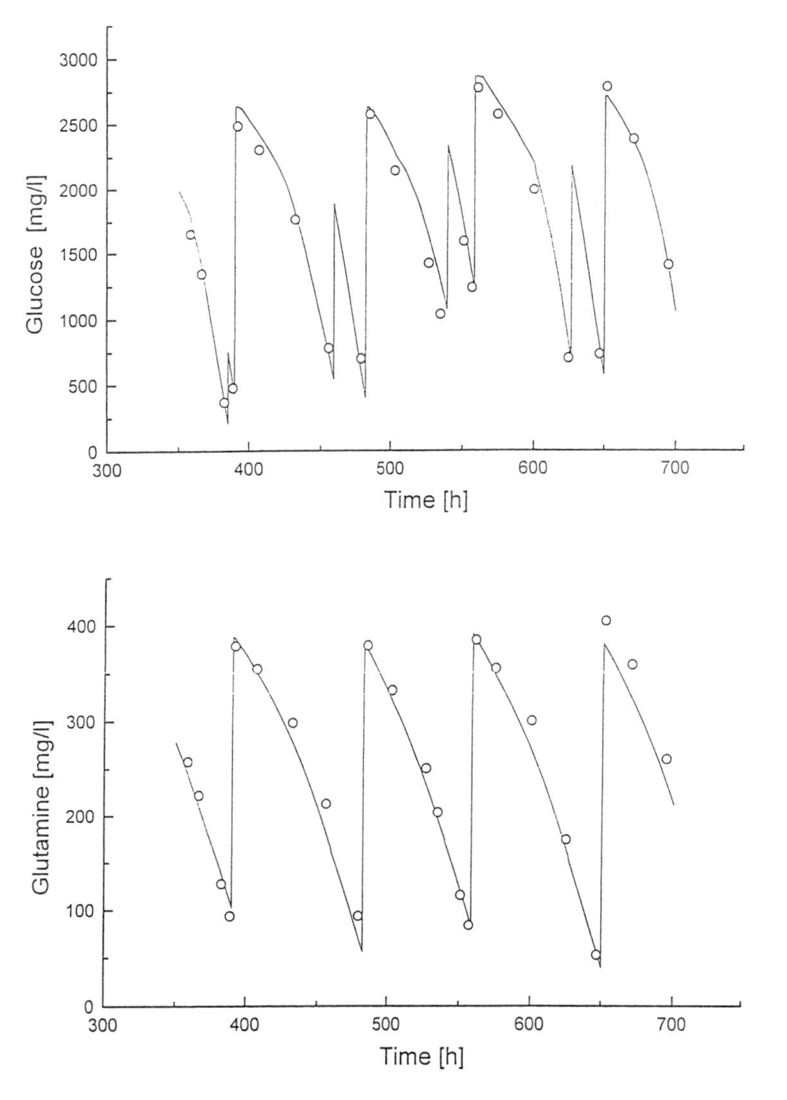

Figure 3a. Estimated signals (curves) together with the measured data (symbols) which became known later.

Continued on next page

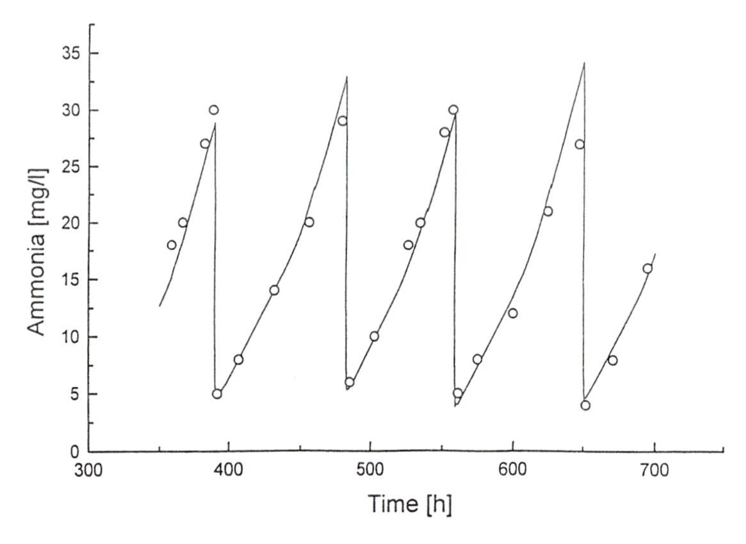

Figure 3a. *Continued*

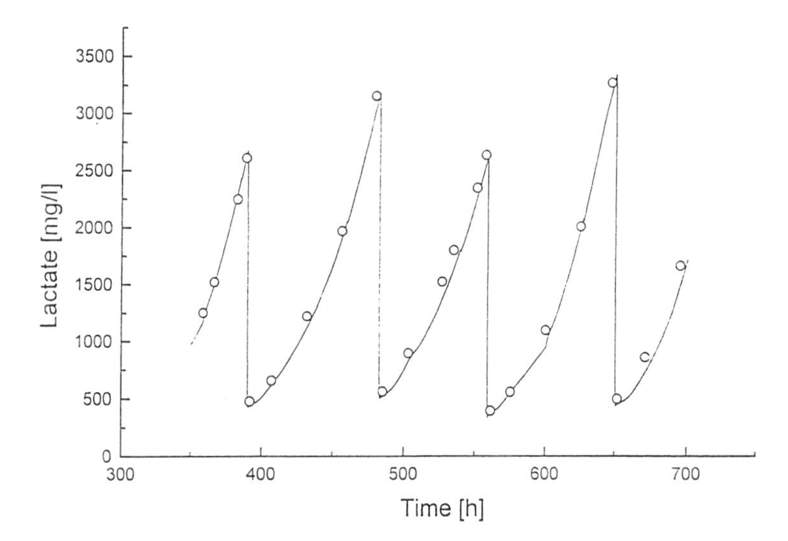

Figure 3b. Estimated signals (curves) together with the measured data (symbols) which became known later.

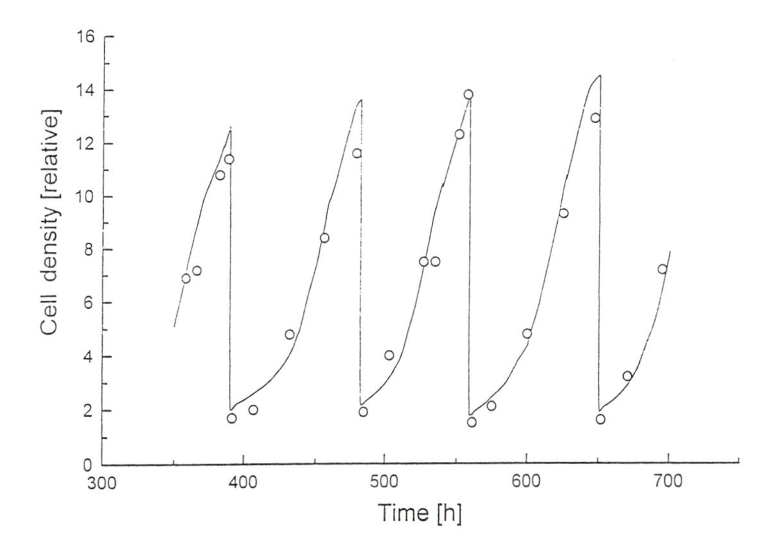

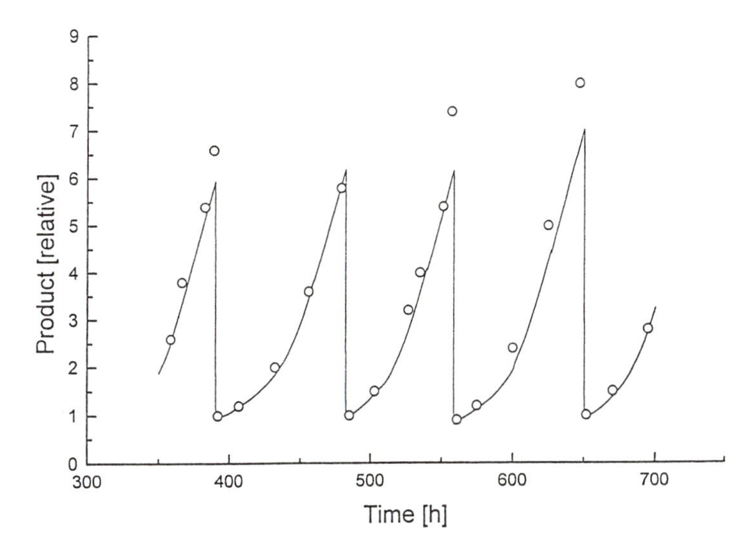

Figure 3b. *Continued*

In order to optimize the feeding, a hybrid model for state prediction was developed. This model is based on a Monod-type correlation supported by an artifical neural network. After adapting the model to the real process the optimal feeding strategy was computed online during the process in an iterative way. Using the computed feeding directions it could be shown that the process runs more stably. Because the feeding strategy was computed online using the actual process data, it was permanently adapted to the process behaviour. In this way, harmful concentrations of the substrate could be avoided and the productivity of the process increased.

Conclusions

The model described combines the available knowledge about a production process of a recombinant protein which is produced in a repeated fed-batch cultivation of mammalian cells. The model has been trained on the available data and gains performance with every production run by using the data to retrain the network component. A consulting software uses that model in order to provide the optimal substrate feeding for given objectives and constraints. It is also possible to suggest a particular feeding profile and to look for the behaviour of the other process parameters involved in the model. Hence, in case of practical constraints, e.g. reduced personnel capacity at the weekends, the process can be adapted accordingly.
The software implemented is capable of learning, hence it will improve its performance with time. A user interface adapted to the process engineer responsible for the production allows to improve and extend the rule system of the heuristic component. The extent of the basic mathematical model clearly depends on the knowledge available about the process under consideration. It should be extended by relationships between the different quantities, e.g., the yield expressions within the balances based on stoichiometric or thermodynamic considerations whenever more detailed knowledge becomes available. Thus, it is strongly suggested to reexamine hybrid models from time to time with the aim of improving the classical mechanistic part of the model.

Literature Cited

1. Lenoard, J.A., Kramer, M.A., Ungar, L.H., A neural network architecture that computes its own reliability, *Comp.Chem.Eng.* **1992**, 16, 819-835
2. Schubert, J., Simutis, R., Dors, M., Havlik, I., Lübbert, A., Bioprocess optimization and control: Application of hybrid modelling, *J. Biotechnol.* **1994**, 35, 51-68
3. Schubert, J., Simutis, R., Dors, M., Havlik, I., Lübbert, A., Hybrid modelling of yeast production processes - Combination of a priori knowledge on different levels of sophistication, *Chem.Eng.Technol.* **1994**, 17, 10-20
4. Xie L. Wang, D.I.C., Fed-batch cultivation of animal cells using different medium design concepts and feeding strategies, *Biotechnol.Bioeng.* **1994**, 43, 1175-1189,

RECEIVED September 15, 1995

Chapter 15

Optimization of an *Escherichia coli* Fed-Batch Fermentation Using a Turbidity Measurement System

Yinliang Chen[1], Alahari Arunakumari[2], Glen Hart[1], and Julia Cino[1]

[1]Research and Development Lab, New Brunswick Scientific Company, Inc., 44 Talmadge Road, Edison, NJ 08818
[1]Enzon Inc., 40 Kingsbridge Road, Piscataway, NJ 08854

A method for optimal on-line control for a fed-batch culture of a genetically engineered E. coli is proposed. AFS-BioCommand, a new software based on windows was used in the process control. The glucose requirement of growth which was calculated from turbidity measurement was compared with the actual feeding. The feeding rate was then adjusted to maintain the designed glucose concentration. Dissolved oxygen was controlled using an agitation cascade with pure oxygen. A temperature induction was used in the process and the fermentation temperature was shifted by the computer control system based on input from on-line turbidity measurements.

The fermentation of recombinant microorganisms for production of large amounts of certain desired proteins requires optimization of cultivation conditions. A microorganism containing a recombinant plasmid is compelled to allocate a fraction of its limited resources for maintenance and replication of the plasmid and for synthesis of plasmid-encoded products. The relationship among cell growth, plasmid content, and expression of plasmid-carried genes are influenced at a fundamental level by cultivation conditions(*1,2*). The presence of foreign genes is known to increase the oxygen requirement of the host. Experiments have shown that the specific growth rates of plasmid-bearing and plasmid-free cells are very sensitive to variations in the dissolved oxygen (DO) level in a particular range. It was observed that a transition in the metabolic pathways may occur when DO levels are shifted beyond a critical level. Furthermore, a dissolved oxygen shock, introduced in the culture by depleting the DO of the bioreactor for a limited period, has been observed to lead to severe plasmid instability for a recombinant *E. coli* strain(*3*).

Most fermentation processes are operated in either batch or fed-batch mode. The fed-batch mode has been introduced in an increasing number of fermentation processes. This is because theoretical analysis as well as experimental data of many fermentation processes have revealed that controlled feeding during a fermentation

0097–6156/95/0613–0155$12.00/0

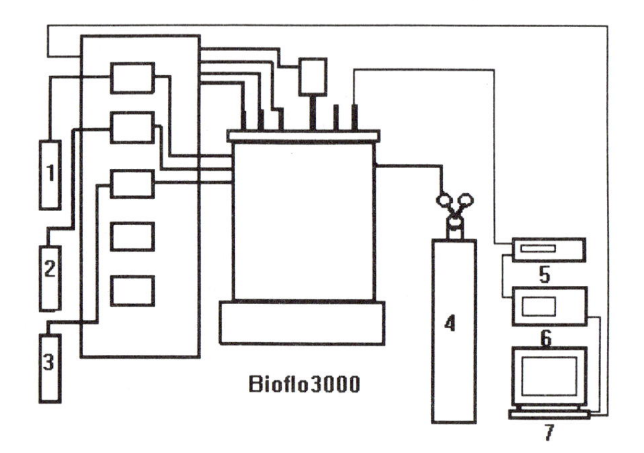

Figure 1 BioFlo 3000 fermenter and on-line control system
1. base solution, 2. antifoam, 3. feeding medium
4. oxygen tank, 5. turbidity measuring system
6. ML4100 multi-loop controller

could effectively overcome such effects as substrate inhibition, catabolite repression, product inhibition, glucose effects, and auxotrophic mutation(*4-6*). Many authors have considered the problem of determining the optimal feed rate for fed-batch fermentation. Aiba and co-workers(*7*) controlled the feed rate of fresh medium based on the respiratory quotient value and Cooney and co-workers used the inlet and outlet oxygen and carbon dioxide concentrations to develop a control algorithm(*8*). Wu and co-workers(*9*) determined the specific growth rate by use of the material balance on oxygen and obtained an optimal feed rate for fed-batch culture based on the moving model. Some researchers(*10*) used the specific growth rate as the control variable and Bentley and Kompala(*11*) used a structured kinetic modeling approach for the process analysis of the recombinant *E. coli* fermentation in these experiments and the instantaneous specific growth rate was directly calculated.

In the present study, a BioFlo 3000 bench-top fermentor connected to a pure oxygen supply was used to meet the high oxygen demand of the process. The glucose requirement of cell growth was calculated via on-line turbidity measuements. AFS-BioCommand, a new software was used to control the process. Optimal glucose feed rates into the fermenter were maintained utilizing regularly measured turbidity data.

Materials and Equipment

Genetically engineered *E. coli* was used in all experiments, which expresses a single-chain Fv(sFv) protein. An expression vector used for sFv expression in *E. coli* contains the hybrid lambda phage promoter O_L/P_R and the *omp*A signal sequence(*12*). The completed sFv expression vectors are transformed into *E. coli* host strain GX6712 which has the mutant temperature-sensitive repressor gene cI^{857} integrated into the chromosome. This provides a transcriptional regulation system where induction of sFv synthesis occurs by raising the culture temperature from 32°C to 42°C.

BioFlo 3000 bench-top fermenters (New Brunswick Scientific Co. Inc.) of 2.5-liter and 5 liter working volumes were used in batch and fed-batch modes with oxygen supplementation in the automatic dissolved oxygen control mode (Fig 1). The turbidity measuring system (Mettler PSC 402, METTLER, Switzerland) was installed to monitor cell mass. The light source is an infrared of a 880 nm wavelength and the receiver consists of a silicone photo diode. Signal transmission via optical fibers is very reliable since it is not subject to any kind of interference. The ML4100 general purpose controller (New Brunswick Scientific Co., Inc.) was used as a communication interface between the turbidity meter and the host computer. The culture medium was prepared and added to the vessel prior to sterilization. 50% glucose solution was used as feeding medium. 6N NaOH solution was used as a base solution.

A frozen vial of *E coli* sample was inoculated into a one liter shaker flask with 200 ml LB medium with ampicillin and incubated at 32°C, 240 rpm for 14 hours in a G20 environment incubator shaker (New Brunswick Scientific Co. Inc.). The entire 200ml inoculum was transferred into the 3.3 liter vessel containing 2 liter medium. The temperature was controlled at 32°C and pH at 7.2.

To determine the dry cell mass, samples were centrifuged and washed once with deionized water followed by dry at 80°C for 60 hours. The optical density was measured at 600 nm with a Perkin-Elmer Lambda 4B UV / VIS Spectrophotometer. Glucose was analyzed using the hexokinase method (Glucose kit 16UV Sigma Chemical Co.) as well as ACCU-CHECK II with test strips (Boehringer Mannheim). The glucose was continuously fed at a certain speed to maintain the glucose level.

The fermentation was controlled and monitored using AFS-BioCommand process control software (New Brunswick Scientific Co., Inc.).The host platform was a 486 computer running DOS and Windows (Microsoft). The software was programmed to implement the algorithm described below using BioCommand control diagrams, a pictorial function-block language. All directly sensed fermentation conditions were reported to the software approximately once per minute. All calculations, including those which controlled glucose addition, and those which determined cell mass and glucose consumption from turbidity measurements were performed on-line twice per minute. The frequent updating of conditions and calculations contributed to precise maintenance of desired glucose and DO levels. The temperature induction was programmed using a BioCommand time profile, a table of time versus temperature setpoint. The entire process was monitored from a computer screen, which displayed the continuously updated fermentor conditions including turbidity values, as well as all calculated values both graphically and numerically.

The On-Line Optimal Algorithm

In this bioprocess, specific kinetics can be expressed in terms of the concentration of the limiting substrate glucose(Glu_X). Therefore, the material balances for cell mass and limiting substrate are sufficient to describe kinetics of the fermentor when operated in fed-batch mode. There are

$$dV / dt = F_{in} - F_{out} \quad (1)$$

$$dX / dt = -(F_{in} / V)* X + \mu X \quad (2)$$

$$dS / dt = F_{in} / V (S_{in} - S_{out}) - \mu X / Y_{X/S} \quad (3)$$

where μ is the specific growth rate, X, the cell mass, V, the working volume of the fermentor and $Y_{X/S}$, is the yield of cell mass with respect to the limiting substrate. Because the feeding rate, F_{in} is very small compared to the working volume and also can be equivalent to sampling volume, F_{out} in this experiment, the generalized balance equations for conversion of substrate (Glu_X) to cell mass (X) can be written as

$$Glu_X = q X \quad (4)$$

where Glu_X is the total glucose [g] consumption, q , the constant which is equal to $1 / Y_{X/S}$ and X, the total biomass [g / l]. Because there is a linear relationship between biomass X and turbidity (turb), the relationship can be expressed as ,

$$X = 0.122 \text{ turb} - 0.73 \quad (5)$$

During the fed-batch fermentation, we can get total glucose feeding Glu_F by integrating the pump flow rate, F (ml / h):

$$Glu_F = c\int F\,dt \quad (6)$$

where c is the glucose concentration of the stock medium (g / ml).

The flow chart of the control program of glucose feeding is given in Figure. 2. This program was executed on a personal computer using the AFS-BioCommand software package.

Results and Discussions

It was found that there is a linear relationship between the turbidity (turb) and the optical density (OD) can be expressed as

$$OD = 0.232 \text{ turb} + 0.667 \quad (7)$$

Another equation then can be obtained after the combination of equation 5 and equation 7, which is used in computer control(Figure 2). In Figure 2, a, b, r, f, K refer to constants which are necessary for the calculation of all parameters. The values of a and b were obtained from equations 4 and 5, and r value was obtained from the calibration of feed pump. f is the multiplication number of feed rate, which can be set by users. K is a constant which is preset feeding rate after the culture temperature is changed from 32°C to 42°C. These values vary depending on such factors as tubing size, pump speed, medium concentration etc. At the start of the fermentation the constants are entered into the software program along with the initial temperature setpoint of 32°C. During the fermentation should the temperature be raised to 42°C or more, the volumetric feed rate of the glucose solution is maintained at the preset constant value K. As long as the temperature are maintained at the growth temperature of 32°C, turbidity measurements are taken from the on-line measurement device and utilized to calculate an average turbidity every six minutes. Since data points are taken by the software every thirty seconds, the sum of the turbidity readings are divided by 12 to determine the average biomass concentration, X. This value X is used to determine the glucose requirement of the culture. The turbidity value for the culture at time 0 is subtracted from the current average turbidity value. Feed totals are then calculated. Using the newly recalculated total for the feed substance, in this case glucose, the value D is calculated, and it is recalculated regularly. If D is greater than or equal to zero, the volumetric feed rate F equals to 0, and if D is less than 0, the feed rate, F, becomes D multiplied by the constant f. The recalculated value is used to recalculate feed total value.

The dry cell weight (DCW) was measured in the fermentation process with pure oxygen supply (See Figure 3) and the relation between DCW and OD is shown

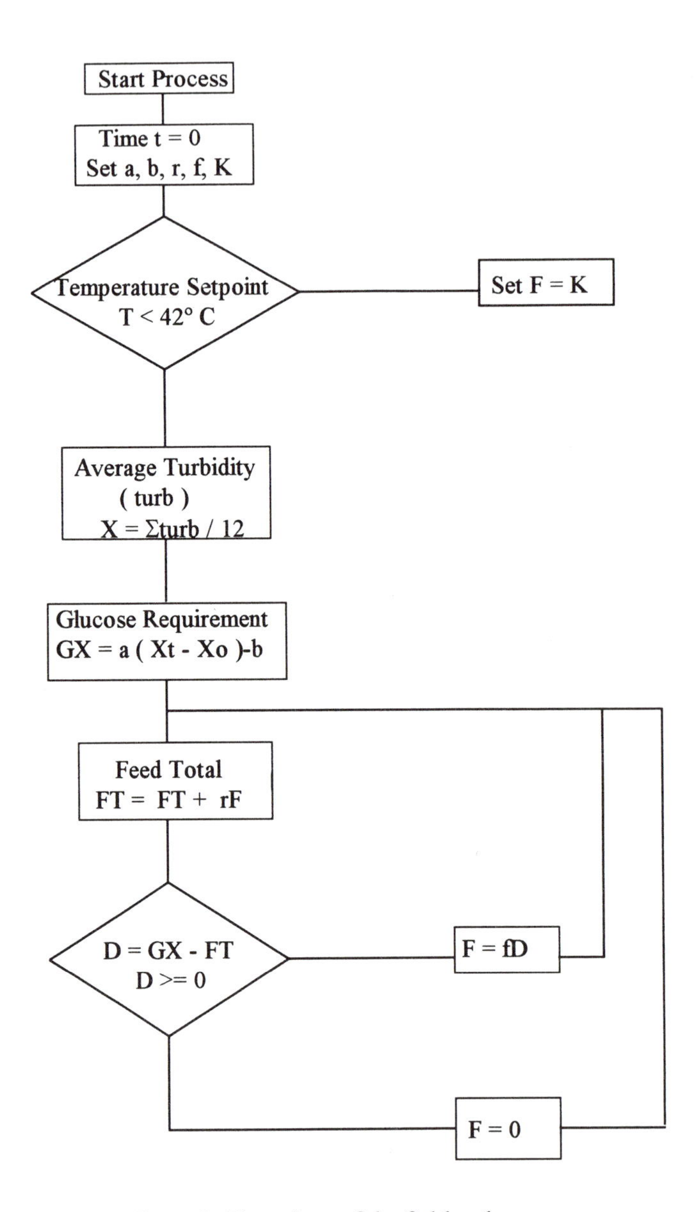

Figure 2 Flow chart of the fed-batch program

on Figure 4. The glucose concentration was analyzed and maintained at a level to meet the demand of the culture as shown on Fig. 5. A temperature induction was employed to induce the protein expression. It was observed that the dissolved oxygen(DO) concentration dropped to zero even before the temperature reached 42°C in fermentations which were not supplemented with oxygen. DO level was maintained during this experiment when the BioFlo3000 fermenter with pure oxygen supplementation was used(Figure 6). Because pure oxygen was used, higher cell density was obtained and the temperature was shifted at an OD value two times higher than without oxygen supply(Figure 7). Both biomass production and the amount of protein expressed were greatly improved.

Conclusion

A method of optimal on-line control for a fed-batch culture of a genetically engineered *E. coli* is proposed. Key elements of the method are oxygen enrichment of the sparge gas, on-line computer control of the fed-batch process, and in-situ turbidity measurements. The fed-batch process with pure oxygen supply greatly increased oxygen transfer rate which could meet the very high oxygen demand of this genetically engineered *E. coli* fermentation. Pure oxygen was supplied automatically when the fermentation temperature was shifted from 32°C to 42°C, so that DO was kept at a very stable level during the entire process. AFS-BioCommand, a new software based on windows was used in the process mathematical modeling and on-line control. The glucose requirement of cell growth which was calculated from turbidity measurement was compared with the actual feeding and then the feeding rate was adjusted for maintaining the designed glucose concentration level. The glucose concentration was controlled at constant level (3 to 4 g/l) under fed-batch process to improve cell growth. The temperature induction was performed at higher cell density in this experiment so that higher productivity was obtained. The turbidity measuring system was used in this fermentation process to detect the biomass directly. It simplified the mathematical modeling in fed-batch control when the AFS-BioCommand computer software and on-line turbidity measurement were used. In this experiment the cell density increased two times more than batch culture and a higher yield of protein product was obtained.

Legend of Symbols

c	glucose concentration of the stock medium (g / L)
F	volumetric feed rate (L / h)
Glu_x	total glucose consumption (g)
OD	optical density
P	product concentration (g / L)
S	substrate concentration (g / L)
t	fermentation time (h)
turb	turbidity reading
V	working volume of a fermenter (L)

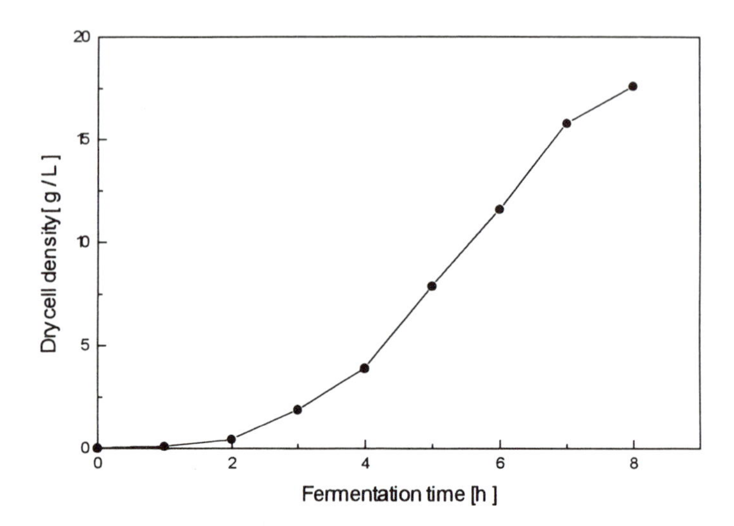

Figure 3. Dry cell weight (DCW) profile in the fermentation process with oxygen supply

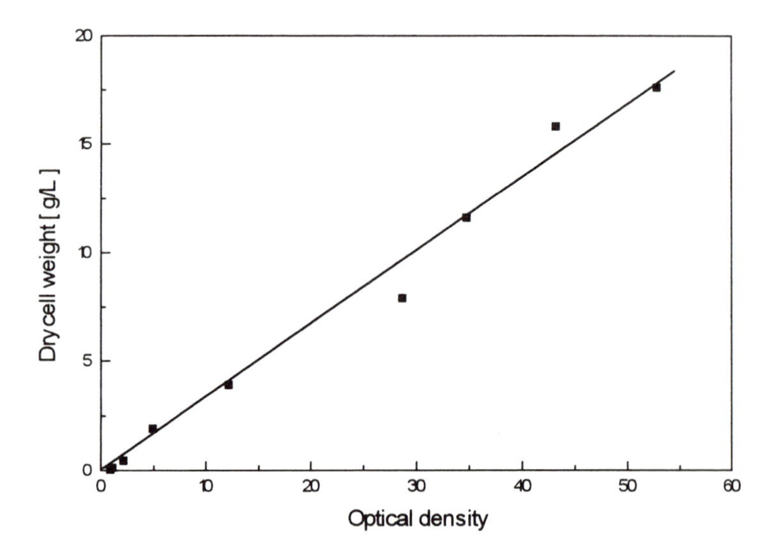

Figure 4. Relation between DCW and OD in the fermentation with oxygen supply

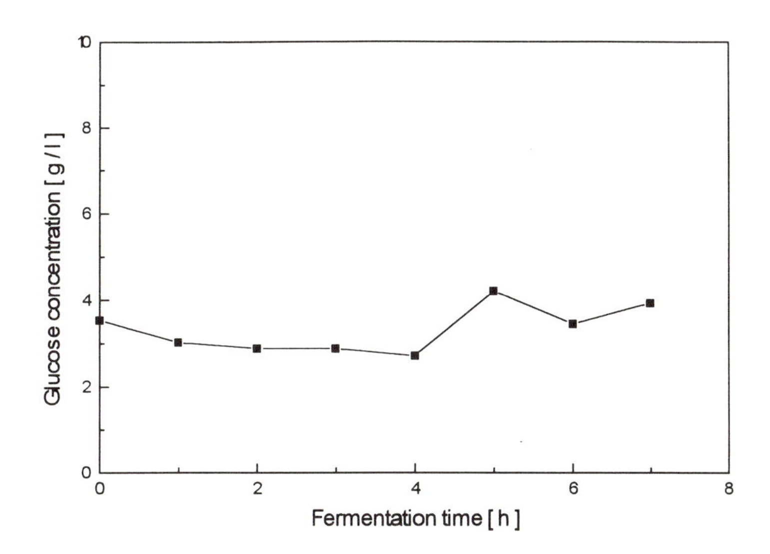

Figure 5 Glucose concentration profile in the fed-batch fermentation process

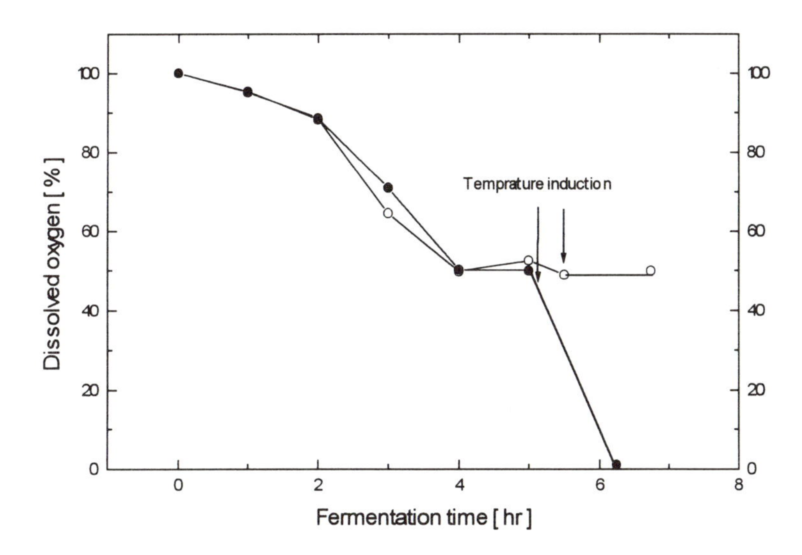

Figure 6. Dissolved oxygen profiles of the fermentations with (o) and without (•) oxygen supply

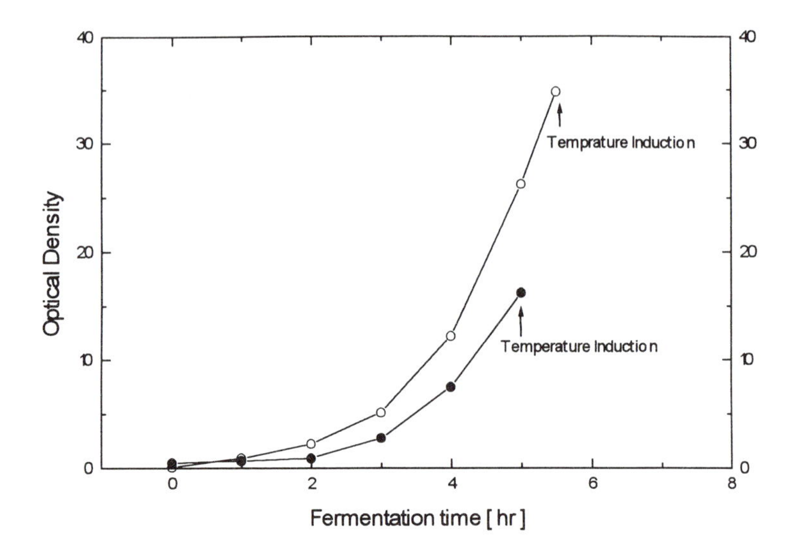

Figure 7. Optical density time profiles in Bioflo 3000 (-o- fed-batch with oxygen supply and -•- batch without pure oxygen supply)

X	biomass concentration (g / L)
y	glucose yield (g / g)
$Y_{X/S}$	yield of cell mass with respect to the limiting substrate
μ	specific growth rate (1 / h)

Literature Cited

1. Flickinger, M.; Rouse, M. P. Biotechnol. Prog. 1993, 9, pp555.
2. Goldwin, D.; Slater, J. H. J. Gen. Microbiol. 1979, 111, pp201.
3. Hopkins, D. J. M.; Betenbaugh, M. J.; Dhurjati, P. Biotechnol. Bioeng. 1987, 29, pp85
4. Agrawal, P.; Koshy, G.; Ramseier, M. Biotechnol. Bioeng., 1989,33, pp115.
5. Modak, J. M.; Lim, H. C. Biotechnol. Bioeng. 1989, 33, pp11.
6. Yang, X., J. Biotechnol. 1992, 23, pp271.
7. Aiba, S.; Nagai, S.; Nishizaqa, Y. Biotechnol. Bioeng. 1976, 18, pp1001.
8. Wei, D., Paurulekar, S. J. and Weigand, W. A., in Multivariable Control of Continuous and Fed-Batch Bioreactors, Biotechnical Engineering VI, Ann. N.Y. Acad. Sci. 1990, Vol 589, pp508.
9. Wu, W.; Chen, K.; Chiou, H. Biotechnol. Bioeng. 1985, 27, pp756.
10. Yamane, T.; Kume, T.; Sada, E.; Takamatsu, T., J. Ferment. Technol., 1977, 55, pp587
11. Bentley, W. E; Kompala, D. S. Biotechnol. Bioeng., 1989, 33, pp49.
12. Whitlow, M.; Filpula, D. Tumour Immunobiology; Oxford University Press; New York, NY, 1993; pp285-287.

RECEIVED July 24, 1995

Chapter 16

The Automation of Two Flow-Injection Immunoassays Using a Flexible Software System

B. Hitzmann[1], B. Schulze[2], M. Reinecke[2], and T. Scheper[2]

[1]Institut für Technische Chemie, Universität Hannover, Callinstrasse 3, 30167 Hannover, Germany
[2]Institut für Biochemie, Abteilung Biochemie, Westfälische Wilhelms-Universität Münster, Wilhelm-Klemm-Strasse 2, 48149 Münster, Germany

In this study two different flow-injection immunoassays are presented as well as the flexible automation system CAFCA (Computer Assisted Flow Control & Analysis), which has been used for their control, uptake measurement, evaluation and visualization. Both immunoassays (a heterogeneous and a homogeneous assay) are based on the principles of flow-injection analysis and were developed for reliable, fast monitoring of relevant proteins in animal cell cultivation processes. Off-line applications of measurements of medium samples as well as on-line application during a mammalian cell cultivation are presented. All results are compared to results obtained with ELISA (Enzyme Linked Immunosorbend Assay). The requirements of the automation of flow-injection immunoassays with respect to their flexible control are discussed.

The highly specific reaction of antibodies with their antigens for the quantification of analyte concentrations is the basis for these flow-injection immunoassays (FIIAs). FIIA is a fast, sensitiv and selectiv assay which can detect a specific protein even in a complex solution. Even in the case where the immunoreaction is slow, FIIA can be applied by carefully controlling injected volume, flow rate, and contact time. This combination has increased the potential for automation of immunoassays dramatically. Among other things the flexibility of reagent addition has led to various FIIA systems based on homogeneous as well as heterogeneous assays. In the homogeneous assay the sample (antigen or antibody) is mixed with a reagent (the immunopartner) applying the merging zone technique. The antigen-antibody complex can than be detected by various techniques utilizing a change in some inherent charac-

0097–6156/95/0613–0165$12.00/0

teristic. To increase sensitivity, the reagent is sometimes labeled. If a detectable change of the compound is not present, the more sensitiv heterogeneous assay can be applied. One of the immunoreactants is immobilized in a reactor which is part of the flow system. Due to the binding of the analyte to the immobilized immunopartner a separation is achieved. The analyte can than be detected, usually after elution. Pollema et al. [*1*], Puchades et al. [*2*] as well as Gübitz and Shellum [*3*] give an overview on the applications of flow injection techniques in immunoanalysis and examples of various immunoanalysis formats. The importance as well as the potential of FIIA for process monitoring are discussed by Degelau et al. [*4*] as well as Mattiasson [*5*].

The application of FIIA systems for the monitoring of real processes – in an industrial environment – will depend strongly on the degree of their automation. Therefore, a flexible automation system is necessary for controlling the valves and evaluating the measurements. If the FIIA system is used for on-line process monitoring, such an automation system has to provide special features for the programming of the course of events so that calibration, measurement, and washing cycles can be processed as required. Until now, only a few software packages for automation have been developed that consider the special requirements of flow-injection analysis [*6 - 10*]. However, the special demands for FIIA system automation are not considered in these systems.

In this contribution two different FIIA systems are presented as well as the automation system CAFCA (Computer Assisted Flow Control & Analysis). The FIIA systems based on a homogeneous assay utilizing a turbidimeter as well as on a heterogeneous assay with an immobilized immunocomponent and a fluorescence detector. The special requirements for their control and their data evaluation procedures will be discussed.

Materials and Methods

The homogeneous FIIA system

For the homogeneous assay a two channel system was applied with one measurement channel as well as one reference channel. As can be seen in Figure 1 the merging zone technique has been used. 50 µl of the sample containing antigen (anti-A-Mab, a monoclonal antibody of the IgG type, Dr. Karl Thomae GmbH, Germany) is mixed with a solution of antibodies (anti mouse IgG, M 7023, Sigma, Germany) diluted in a buffer, which has also been used as the carrier (0.01 M sodium phosphate, 0.077 M NaCl, pH=7.4, 3 % polyethylene glycol 6000). Peristaltic pumps (Ismatec IPS 4) are used to achieve a pulsation free transport of fluids. Passing the reaction coil where the immunoreactants were incubated at 37°C for 0.2 min. (stopped flow) the turbidity was measured at 340 nm. For the reference channel no immunoreactant was supplied to get the medium blank absorption. The difference of peak height of both channels can be used with a linear regression model to calculate concentrations as will be explained below in more detail. To clean the tubes a 3 % mucasol solution (Merz+Co, Germany) has been used.

The heterogeneous FIIA system

A heterogeneous immunoassay was developed to monitor the IgG concentration during a fermentation run. The assay is based on the highly selective binding reaction between Protein G and IgG and, in contrast to the previously described assay, two phases are involved. In the solid phase, Protein G is immobilized on Sepharose 4 B (Pharmacia) contained in a 0.7 ml cartridge. The advantage of this design is that the expensive Protein G is reused. The liquid phase is a fermentation broth containing IgG which is introduced into the column. As can be seen in Figure 2 the analysis can be subdivided into four major steps. The sample is first injected (injection volume 40 μl, injector by Knauer) into the carrier buffer (0.1 m glycine, pH 6.6, 55 mM NaCl, flow rate 1 ml/min). Once the sample flows into the cartridge, the analyte binds with the immobilized Protein G (panel 1) which is present in excess. Next the cartridge is rinsed with the carrier buffer in order to remove all unbound substances (panel 2). Finally, the IgG is eluted (panel 3) using a pH shift (0.1 m glycine, pH 2, flow rate 1 ml/min). The analyte in the effluent is detected using a on-line fluorimeter by measuring the protein fluorescence (280/340 nm, Merck-Hitachi, model F1030). Thereafter the cartridge is equilibrated (panel 4) by the carrier buffer and the next sample can be injected. For sample injection the FIA system is coupled with a autosampler (Isis, Isco, Germany). The equipment is described in detail elsewhere [*11*]. To ensure that no growth of microorganisms occurs the tubing from the autosampler to the FIA was washed after each sample. One sample may be analyzed every six minutes with an injection period of 10 seconds, a washing period of 170 sec, an elution period of 150 sec and an equilibration period of 30 sec.

The computer system

Using Turbo Pascal 7.0 and a toolbox for real time application (RTKernel V4.0, On Time Informatik GmbH, Germany) the automation system CAFCA (Computer Assisted Flow Control & Analysis, ANASYSCON GmbH, Germany) was developed. It runs on MS-DOS-PCs type 80386 or higher with at least 1 mega byte of memory as well as a hard disk with at least 1 mega byte for software and enough space for storing the measurement data. CAFCA supports commercially available interface cards (such as from the PC-LabCard series) for A/D conversion and I/O control. It was used not only for the automation of the FIIA systems but also to control the auto-sampler. Therefore, all the usual hardware of flow analysis systems is supported.

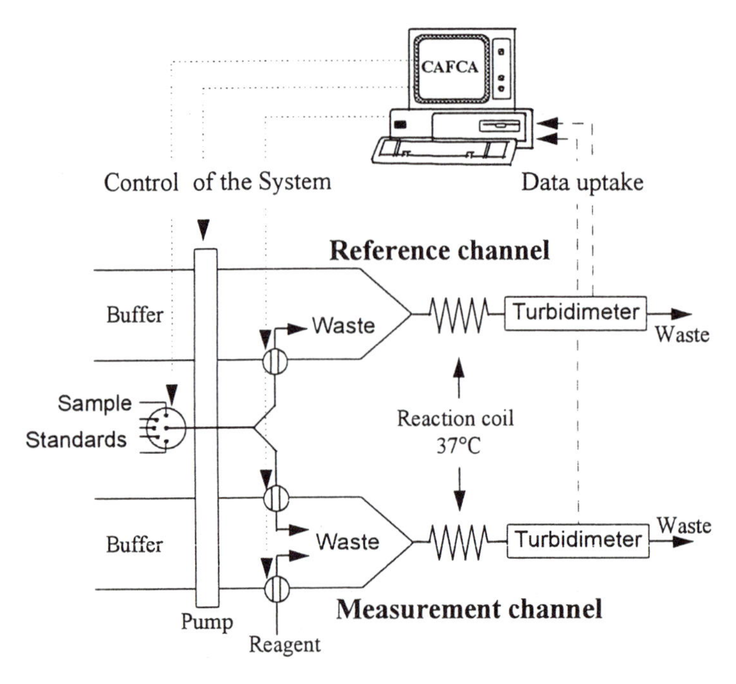

Figure 1. A schematic diagram of the homogeneous FIIA system including the automation system CAFCA used for the monitoring of a mammalian cell cultivation

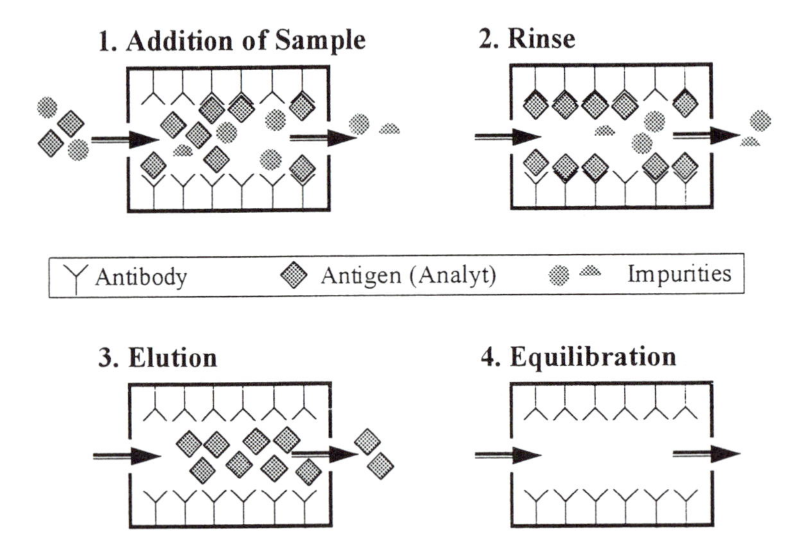

Figure 2. The four major analysis steps of the heterogenous FIIA system

Results and Discussion

CAFCA was developed to serve as a universal tool for automation of flow systems. The main features covered by CAFCA are the configuration, the programming of the course of events, the calibration, the on-line procedures (control, measurement, and evaluation), and the off-line evaluation as presented in Figure 3. Each of these features is implemented in a separate module (execution file), to be more flexible and to avoid problems with the constrained memory of MS-DOS. CAFCA can be adapted to a specific FIIA system by the configuration module. The number of measurement or reference channels, the evaluation method for each channel (for example, peak height, area or width), or the measurement uptake rate for the recorded signals are selected with this module. To determine the control of the valves as well as the pumps one can apply the module to program the course of events. Using this module, the time of incubation (stopped-flow method) is defined as well as the time for washing, elution, reequilibration, calibration or measurement cycles. After the definition of these cycles they can be combined in an arbitrary manner. Various loops of these cycles can be programmed containing other loops. This kind of programming guarantees a very flexible control of FIIA systems as will be described below. In addition, during development of the FIIA systems it has become obvious, that optimal performance requires fine tuning of the operation conditions. Programs can easily be written or changed for this purpose. Programming the course of FIIA systems requires more flexibility than for usual flow injection systems because additional cycles are necessary, such as an elution cycle or a reequilibration cycle for a heterogeneous assay. Furthermore, when using a FIIA system for on-line purposes, calibration cycles have to be performed in either a cyclic or automated fashion or on command of the operator manually. Using CAFCA both types of calibration can be performed.

To specify the calibration, i. e. the number of standards as well as the calibration model, a special module has been developed as is shown in Figure 3. Another module, which serves for the on-line procedures, performs the control of the FIIA system as well as the measurement uptake and evaluation. The results can be transferred via a serial port or an Ethernet interface to a process data management system. Using CAFCA all data are visualized and stored so that an inspection and a pretreatment of the evaluation is always possible. The latter can be carried out by using the off-line evaluation module. CAFCA has been used for the automation of various flow systems. The application to the homogeneous and heterogeneous FIIA is described now in more detail focusing on the control of the FIIA operation as well as on a reliable evaluation procedure, respectively.

Computer Assisted Flow Control & Analysis

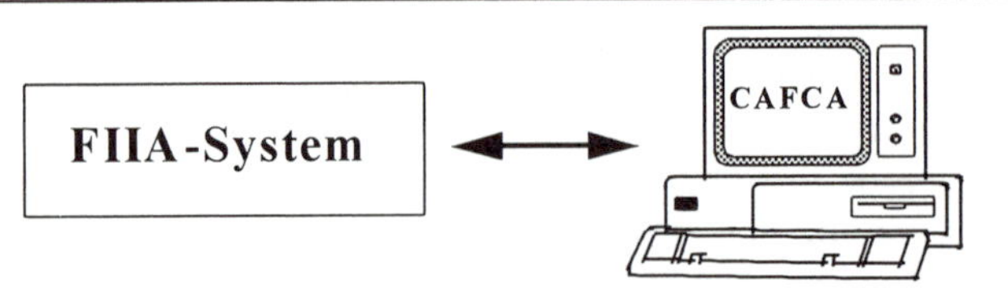

Configuration
- adapting CAFCA to a specific flow system
- adapting CAFCA to the computer hardware
- selecting the best evaluation procedure

Programming the course of events
- programming the valves
- programming the pumps
- programming various cycles

Calibration
- specifying the number of standards
- specifying the calibration models

On-line procedures
- control
- measurement uptake
- evaluation
- visualization
- data storage

Off-line evaluation
- reevaluation
- inspection of signals

Figure 3. A schematic of the main features of CAFCA.

Using CAFCA for the control of the homogeneous FIIA system

For the measurement of anti-A-Mab during the cultivation of mammalian cells the homogeneous assay was utilized. As programmed by using CAFCA the course of a whole measurement cycle for the determination of each analyte concentration is presented in Figure 4. It can be seen that one concentration value is based on the measurement of three individual samples following by an averaging procedure. The calculated standard deviation can be used for on-line validation purposes to ensure reliable monitoring. The whole analysis of a triple measurement took about 16 min. Since the product formation of mammalian cells is slow, this procedure ensures a more reliable determination of analyte concentration. Before this FIIA system was used for on-line measurements, it was exhaustively validated off-line as described by Schulze et al. [*12*].

To ensure that the FIIA system runs fully automated during cultivation the course of events was programmed as presented in Figure 5. Before measurement the system performs a calibration cycle using three different standards. A measurement cycle is then performed followed by a washing cycle. During cultivation these two cycles were processed until the FIIA system was stopped manually. The washing cycle was introduced to reduce the consumption of reagents as well as to prevent microbial growth in the tubing since the duration of monitoring was about 140 h. During the monitoring phase no automatic recalibration was programmed, since this could also be activated on-line just by pressing a function key. In Figure 6, the cultivation profile is presented as measured on-line by FIIA as well as off-line by ELISA. The two measurements agree very well with only one calibration each day. The mean standard deviation of FIIA measurements is 4.5 % with respect to the triple analysis. Applying a fourth order regression model of FIIA measurements, which is also shown in Figure 6, the average relative error of anti-A-Mab determination is 2.9 %. The average relative error of ELISA with respect to the regression model is 7.0 %. Therefore, a good agreement between these two measurements exists. Comparing FIIA with ELISA, which is still the widely used assay for protein monitoring, the latter is more labor intensive and time consuming. Applying FIIA the data can be used for true on-line monitoring which is not realistic for ELISA. Furthermore, the degree of automation is much higher for FIIA than for ELISA, through the use of a flexible automation system like CAFCA. The automation of FIIA is not only based on the ability to control valves and pumps in any order but also to process the recorded signal reliably, i. e. to calculate the analyte concentration. This holds especially in an industrial environment, where electronic noise is present everywhere and could potentially interfere with measurements. If the concentration values are used for control purposes then their reliability is even more important. For this reason, we have developed an expert system for the permanent supervision of a flow injection analysis system [*13*].

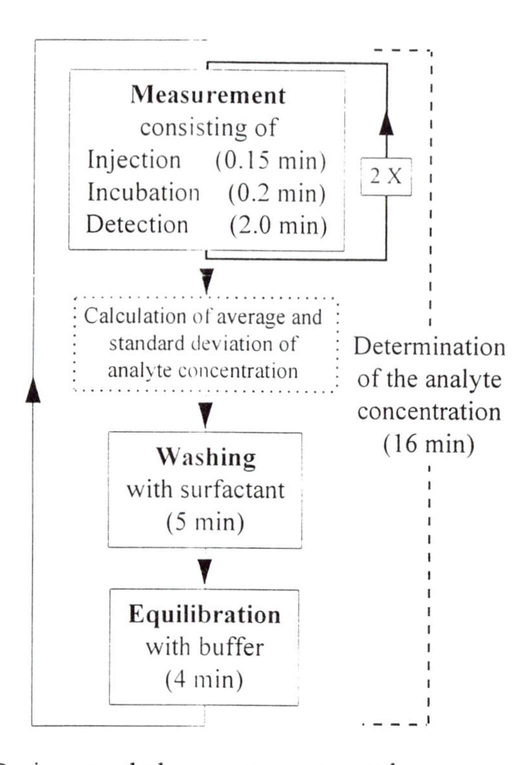

Figure 4. During a whole measurement cycle as programmed by CAFCA different activities can be carried out such as: Measurement, calculation of analyte concentration, washing, and equilibration

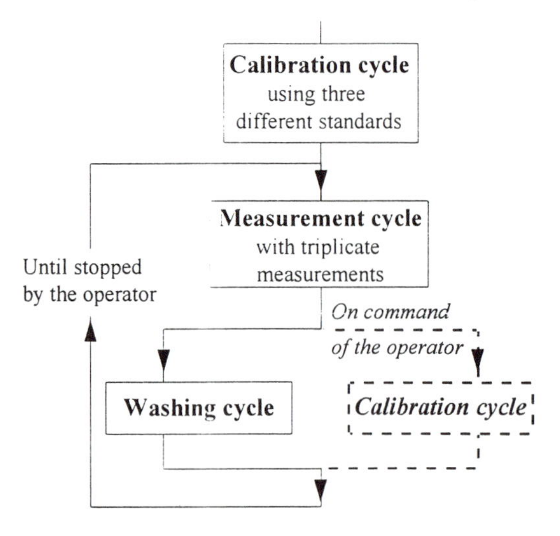

Figure 5. To guarantee a fully automated operation during the mammalian cell cultivation the course of events has been preprogrammed by linking various cycles together.

Using CAFCA for the control of the heterogenous FIIA system

The heterogeneous FIA system was fully controlled by a PC operating the CAFCA software. Pumps, valves and an autosampler for sample loading are controlled by the software. The signal from the fluorescence detector is processed by CAFCA as well. Due to a separation between the analyte and impurities, the medium does not disturb the assay, so a second channel for the media is not required. The peaks of one analysis are stored in two files, one for the total protein concentration (injection cycle panel a and b) and one for the IgG concentration (elution cycle panel c and d). The elution peak area, integrated by the CAFCA program, was determined to measure the sample concentration. For greater reliability measurements were taken in triplicate. This leads to six cycles for every sample, three injection and three elution cycles. The assay was calibrated on a daily basis using IgG standards present in fermentation media. Due to a decline in the binding efficiency of the cartridge, the slope of the calibration curve became smaller over long times. The calibration curves were always linear by a least squares analysis with correlation coefficients ranging from 0.998 and 0.9998. The linearity did not to depend on the media composition. Using the system, several fermentations were monitored at 24 hours sampling intervals. Cell free samples were taken from the reactor and injected in the FIA system. The results of the FIA assay are compared to those from the ELISA (Figure 7). The median difference between the two analysis as determined from 48 samples was 6.2%. Reproducibility of the FIA assay was checked by analyzing each sample three times and the mean standard variation was 3.6%. The data show that a FIA system combined with a control and data acquisition software leads to a precise stand alone measurement device with high accuracy.

Conclusion

In this contribution the automation system CAFCA and its application to a homogeneous as well as a heterogeneous FIIA system is presented. Using these FIIA systems the ability of CAFCA to control valves and pumps in a cyclic manner is discussed. Applying CAFCA various cycles such as washing, reequilibration, calibration, and measurement cycles can be defined and processed in different loops. CAFCA can be programmed to control a FIIA system during an entire whole process as demonstrated for a mammalian cell cultivation. Using a flexible automation system for FIIA the application of these analyzers for industrial purposes will increase significantly. On the one hand it will increase the capital investment costs, but on the other hand it will reduce labor costs.

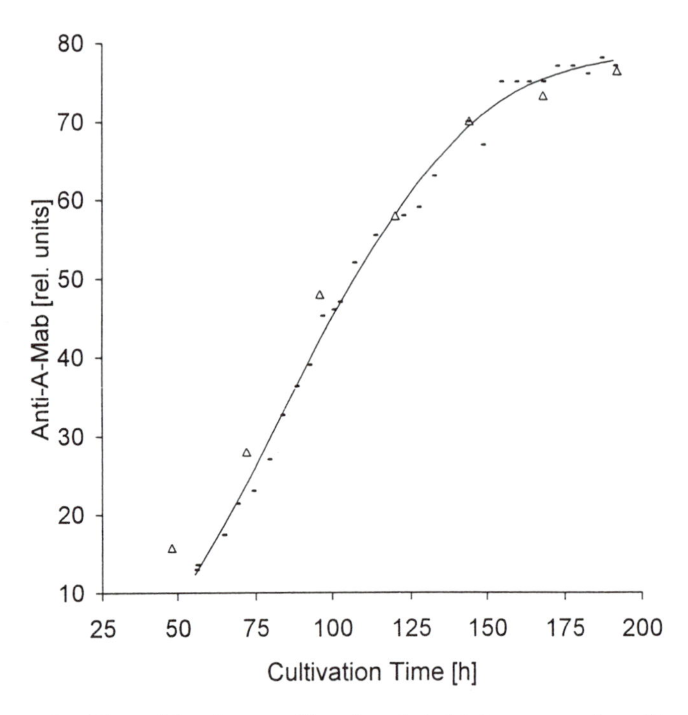

Figure 6. The cultivation profile of anti-A-Mab measured on-line by FIIA (-) and off-line by ELISA (Δ) as well as the regression model of the FIIA measurement (solid line) during a mammalian cell cultivation.

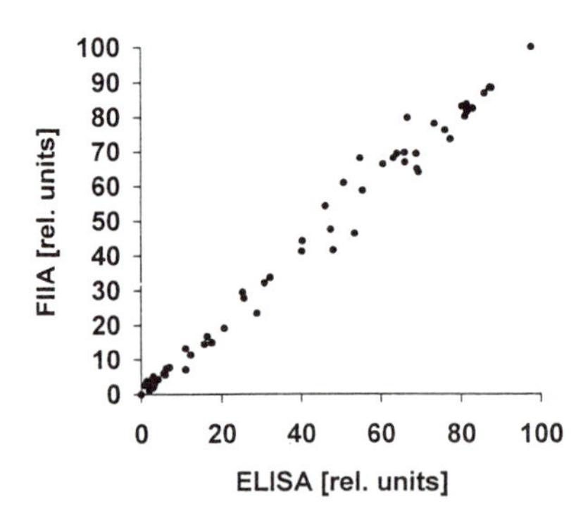

Figure 7. A comparison of measurements optained with the ELISA system and the heterogenous FIIA system

Literature Cited

1. Pollema, Cy H.; Ruzicka, J.; Lernmark, Å.; Christian, G.D. *Microchem. J.* 45(1992)121
2. Puchades, R.; Maquieria, A.; Atienza, J.; Montoya, A. *Crit. Rev. Anal. Chem.* 23,4(1992)301
3. Gübitz, G.; Shellum, C. *Anal. Chim. Acta* 283(1993)421
4. Degelau, A.; Freitag, R.; Linz, F.; Middendorf, C.; Scheper, T.; Bley, T.; Müller, S.; Stoll, P.; Reardon, K.F. *J. Biotechnol.* 25(1992)115
5. Mattiasson, B. Flow injection bioanalysis - a convenient tool in process monitoring and control, Proceedings of the 6th European Congress on Biotechnology, Florence, 1993, p. 869-872
6. Clark, G.D.; Christian, G.D.; Ruzicka, J.; Anderson, G.F.; van Zee, J.A. *Anal. Instr.* 18(1989)1
7. Busch, M.; Polster, J. *Trends Anal. Chem.* 11(1992)230
8. Marshall, G.D.; van Staden, J.F. *Anal. Instr.* 20(1992)79
9. Busch, M.; Höbel, H.; Polster, J. *J. of Biotechnol.* 31(1993)327
10. Spohn, U.; van der Pol, J.; Eberhardt, R.; Joksch, B.; Wandrey, Ch. *Anal. Chim. Acta* 292(1994)281
11. M. Reinecke, FIA-Bioprozessanalytik, VDI Verlag, Düsseldorf, Reihe 17, Nr. 119
12. Schulze, B.; Middendorf, C.; Reinecke, M.; Scheper, T.; Noé, W.; Howaldt, M. *Cytology* 15(1994)259
13. Hitzmann, B.; Gomersall, R.; Brandt, J.; van Putten, A. (1995) An expert system for the supervision of a multi channel flow injection analysis system. In: Recent Advances in Biosensors, Bioprocess Monitoring, and Bioprocess Control, K.R. Rogers, A. Mulchandani, W. Zhou, Eds., ACS Symposium Series, American Chemical Society, Washington, D.C. (this vol.)

RECEIVED August 23, 1995

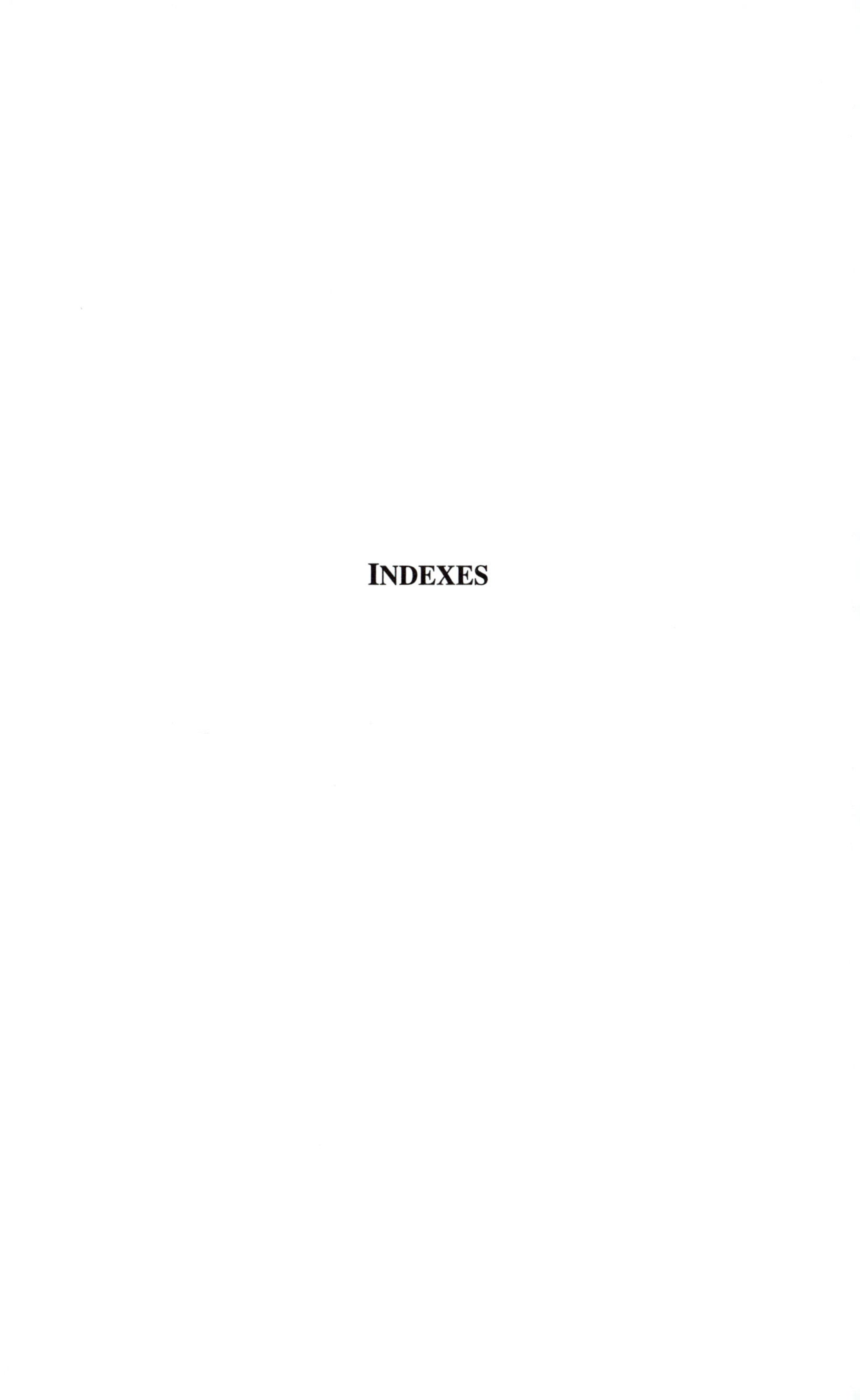

Indexes

Author Index

Arnold, Mark A., 116
Arunakumari, Alahari, 155
Baer, Richard L., 9
Bambot, Shabbir B., 99
Barrows, Lisa C., 61
Bauer, Christian, 70
Bier, Frank F., 70
Brandt, J., 133
Carter, Gary, 99
Chen, Yinliang, 155
Chung, Hoeil, 116
Cino, Julia, 155
Doherty, Thomas P., 9
Dors, M., 144
Eldefrawi, Amira, 19
Eldefrawi, Mohyee, 19
Emanuel, Peter, 19
Förster, Eva, 70
Ghindilis, Andrey L., 70
Gomersall, R., 133
Gueguetchkeri, Manana, 44
Gupta, Adarsh, 44
Hart, Glen, 155
Hitzmann, B., 133,165
Horvath, John J., 44
Kaden, H., 70
Kinnear, K. T., 82
Lakowicz, Joseph R., 99
Ligler, Frances S., 33
Lübbert, A., 144
Makower, Alexander, 70
Meier, Helmut, 110
Michael, Norbert, 70
Micheel, Burkhard, 70
Monbouquette, H. G., 82
Mulchandani, Ashok, 2,61,88
Murhammer, David W., 116
Pease, Mark D., 33
Penumatchu, Devi, 44
Pfeiffer, Dorothea, 70
Rao, Govind, 99
Reinecke, M., 165
Rhiel, Martin, 116
Rogers, Kim R., 2,19
Scheller, Frieder W., 70
Scheper, T., 165
Schulze, B., 165
Shriver-Lake, Lisa, 33
Simutis, R., 144
Sipior, Jeffrey, 99
Spira-Solomon, Darlene, 9
Szeponik, Jan, 70
Tom-Moy, May, 9
Tran-Minh, Canh, 110
Valdes, James, 19
van Putten, A., 133
Weetall, Howard H., 44
Wollenberger, Ulla, 70
Wright, Jeremy, 19
Zhou, Weichang, 88
Zhou, Xiangji, 116

Affiliation Index

BST Bio Sensor Technology GmbH Berlin (Germany), 70
Centre Science des Processus Industriels et Naturels/Biotechnology (France), 110
Enzon Inc., 155
Hewlett-Packard Company, 9
Kurt-Schwabe-Institute for Measuring and Sensor Technology (Germany), 70
Max-Delbrück-Center of Molecular Medicine (Germany), 70

Merck & Co., Inc., 88
National Institute of Standards and Technology, 44
Naval Research Laboratory, 33
New Brunswick Scientific Company, Inc., 155
Research Center of Molecular Diagnostics and Therapy (Russia), 70
Research Institute of Molecular Pharmacology (Germany), 70
U.S. Army Edgewood Research, Development, and Engineering Center, 19
U.S. Environmental Protection Agency, 2,19
Universität Hannover (Germany), 133,144,165
University of California—Los Angeles, 82
University of California—Riverside, 2,61,88
University of Iowa, 116
University of Maryland—Baltimore, 99
University of Maryland School of Medicine, 19
University of Maryland School of Pharmacy, 19
University of Maryland—Baltimore County, 99
University of Potsdam (Germany), 70
Westfälische Wilhelms-Universität Münster (Germany), 165

Subject Index

A

Acoustic wave device as LC detector
  amount of analyte in detector, 15,16*f*
  chemical immobilization procedure, 13
  chromatograms of blank and human immunoglobulin G samples, 13–15
  chromatographic system, 11
  continuous measurement ability, 15,17
  experimental description, 11,13
  limitations, 17
  sensorgrams, 14*f*,15
  surface acoustic wave devices and electronics, 11,12*f*
Acridine orange
  DNA binding, 45
  use in DNA intercalation, 44–58
Adrenal chromaffin cells, catecholamine detection, 75,77,78*f*
Advanced process state estimation, prediction, and control via hybrid process modeling, 144–154
Aminophenols, use as sensor for subnanomolar concentrations, 70–80
Analyte concentrations, measurement using fluorescent lifetime, 99–108
Analytical instrumentation, advances, 88–94
Aniline, detection and measurement by DNA intercalation, 44–58
Anisotropy, calculation, 48
Anthracene, detection and measurement by DNA intercalation, 44–58
Antibody
  cloning, 29–31
  measurement of affinity for antigen, 20*f*,22
Antibody affinities, detection using immunosensors, 19–31
Antigen, measurement of antibody affinity, 20*f*,22
Aqueous media, glutamine and asparagine selective measurement by near-IR spectroscopy, 116–130
Ascorbic acid, role in electroenzymatic sensing of fructose using fructose dehydrogenase immobilized in self-assembled monolayer on gold, 85,86*f*
Asparagine selective measurement in aqueous media by near-IR spectroscopy
  absorption bands, 118–119
  apparatus, 117
  calibration models, 123,128–130

Asparagine selective measurement in aqueous media by near-IR spectroscopy—*Continued*
experimental description, 117,118
Fourier filtering effect on partial least-squares regression, 122–124*t*,127*f*
individual absorption spectra, 117,120*f*
partial least-squares regression, 119–122,124–126
Assay configurations for environmental monitoring
competitive binding assay, 35–37
sandwich assay, 35
Automation, flow injection immunoassay system using flexible software system, 165–174

B

Benzanthracene, detection and measurement by DNA intercalation, 44–58
Benzidine, detection and measurement by DNA intercalation, 44–58
Benzofluoranthenes, detection and measurement by DNA intercalation, 44–58
Benzo[*a*]pyrene, detection and measurement by DNA intercalation, 44–58
Bioaffinity receptors, examples, 3
Biocatalytic receptors, examples, 3
Biological receptors, types, 3
Biologically important species, spectral features for measurement selectivity, 117
Biomolecular recognition
applications, 2
biosensors, 3
concept, 3
environmental monitoring, 5–6
function, 2
future directions, 6–7
process control monitoring, 4–5
requirements, 2–3
Bioprocess and clinical analytes using lifetime-based phase fluorometry, optical measurement, 99–108
Bioprocess chromatography, detector requirements, 9–10
Bioprocess monitoring and control advances
biomolecular recognition, 4–5
components, 89
control concepts, 93–94
control strategy criteria, 88–89
expert systems, 89
future challenges, 94
growth and development, 88
on-line measurements
flow injection analysis, 91–92
high-performance LC, 91–92
image analysis, 90–91
off-gas analysis, 92–93
optical density probes, 89–90
Biosensor(s)
advances, 88–94
advantages, 3
antibody selectivity for analyte, 25–29
criteria, 3
detection limits, 22,25
development, 110
environmental monitoring, 5–6
fiber-optic environmental monitoring, *See* Fiber-optic biosensor use in environmental monitoring
future directions, 6–7
measurement of antibody affinity for antigen, 20*f*,22
on-line penicillin monitoring during production by fermentation
automation procedure, 111–112
calibration curve, 113
comparison to high-performance LC method, 113,114*f*
detection cell, 111,112*f*
flow injection system, 112
importance of measurement, 113
penicillin concentration vs. fermentation time, 113,114*f*
response time, 114
stability, 114
process control monitoring, 4–5
regeneration, 24–26*f*
sensitivity, 22
specificity, 22,23*f*

Biosensor probes
formation, 34
sensitivity, 35
specificity, 34
Burkholderia cepacia G4 5223–PR1, detection, 37–40

C

Carbon dioxide production rate, on-line off-gas analysis for determination, 92–93
Carbon dioxide sensor, use of lifetime-based fluorometry, 103,105–107
Carcinogen detection and measurement by DNA intercalation, polyaromatic, *See* Polyaromatic carcinogen detection and measurement by DNA intercalation
Catecholamines
detection from adrenal chromaffin cells, 75,77,78*f*
use as sensor for subnanomolar concentrations, 70–80
Cell cascade like sequential activation of enzymes, signal amplification, 70
Chemical mixtures
cross-reactivity of constituents, 25
immunosensor detection, 19–31
Chemically modified electrode for hydrogen peroxide measurement by reduction at low potential
analytical characterization, 66,67*f*
application, 67–68
construction, 62–63
effect of number of electrochemical polymerization scans, 63,64*f*
effect of pH, 63,65–66
experimental description, 62–63
hydrodynamic voltammetry, 63,64*f*
interference, 66–67
Clinical analytes, optical measurement using lifetime-based phase fluorometry, 99–108
Cloning
antibodies, hybridoma technology, 29–31
detection using immunosensors, 19–31
Cocaine detection
biosensors, 25,27
enzyme sensors, 80
Coenzyme Q_6, electroenzymatic sensing of fructose, 82–86
Competitive binding assay
advantages, 35
groundwater contaminant detection, 40–41
procedure, 37
Computer-assisted flow control and analysis (CAFCA) system
calibration, 169,170*f*
description, 167
on-line procedures, 169,170*f*
programming, 169,170*f*
Computer technology, advances, 88–94
Control of bioprocesses
advances, 88–94
monitoring using biomolecular recognition, 4–5
Critical angle, determination, 34
Cross-reactivity of constituents in chemical mixture, measurement, 25
Cultivation conditions, optimization for fermentation of recombinant microorganisms, 155

D

Detection limit
definition, 22
immunosensor, 22,25
Detectors for LC, types, 9
4,6-Diamidino-2-phenylindole chloride, 44–58
Dibenzanthracene, detection and measurement by DNA intercalation, 44–58
DNA intercalation, polyaromatic carcinogen detection and measurement, 44–58
Drift, role in expert system for multichannel flow injection analysis system supervision, 139,140*f*
Dyes, intercalative interaction with DNA for polyaromatic carcinogen detection and measurement, 44–58

E

Electrochemical sensors
 alternative of fluorescence detection, 100
 problems, 99–101
Electrode for hydrogen peroxide measurement by reduction at low potential, chemically modified, *See* Chemically modified electrode for hydrogen peroxide measurement by reduction at low potential
Electroenzymatic sensing of fructose using fructose dehydrogenase immobilized in self-assembled monolayer on gold
 calibration curves, 84*f*,85
 control voltammograms, 84*f*,85
 effect of ascorbic acid on current response, 85,86*f*
 experimental description, 82,83
 stability, 86
 time dependence of current response, 85
Environmental monitoring
 assay configurations, 35,37
 biomolecular recognition, 5
 fiber-optic biosensor, 33–41
 scope, 33
Enzyme-based biosensors, development as continuous monitors for bioreactors, 116–117
Enzyme electrodes based on hydrogen peroxide detection
 approaches, 61–62
 description, 61
Enzyme-linked immunosorbent assays, applications, 19
Enzyme sensors
 development, 110
 penicillin monitoring, 110–114
 subnanomolar concentrations
  catecholamine detection from adrenal chromaffin cells, 75,77,78*f*
  cocaine detection, 80
  effect of concentration, 71,74*f*
  effect of glucose, 71,73*f*
  efficiency, 75
  experimental description, 71
Enzyme sensors—*Continued*
  functions, 75
  immunoreaction indication, 77
  norepinephine detection, 75,76*f*
  principle of signal amplification, 71,72*f*
  redox label immunoassay, 80
  sandwich immunoassay for immunoglobulin G, 77,79*f*
  sensitivity, 71,75,76*f*
Equilibrium constant of inhibition, measure of cross-reactivity, 25,27*t*
Escherichia coli fed-batch fermentation optimization using turbidity measurement system
 apparatus, 156*f*,157
 dissolved oxygen profiles, 161,163*f*
 dry cell weight, 159,162*f*
 dry cell weight–optical weight relationship, 159,161,162*f*
 experimental description, 157–158
 flow chart, 159,160*f*
 glucose concentration profile, 161,163*f*
 on-line optimal algorithm, 158–159
 optical density–time relationship, 161,164*f*
 previous studies, 155,157
 turbidity–optical density relationship, 159,160*f*
Ethidium bromide, use in DNA intercalation, 44–58
Evanescent wave, description, 34
Expert system for multichannel flow injection analysis system surpervision
 comparison to single-channel system, 139–141
 evolution of residence time, half-width, and drift during occurrence of fault, 139,140*f*
 experimental description, 135
 faults, 136*t*,137,141–142
 features used to recognize faults, 137,138*t*
 measurement signals during breakdown of carrier stream, 137–139
 multichannel flow injection analysis system structure, 134*f*,135
 structure, 135–137

Expert system for multichannel flow injection analysis system surpervision—*Continued*
undisturbed and disturbed signal
glucose channel, 139–141
protease stopped-flow channel, 141,142*f*
Explosives, detection, 40–41

F

Fed-batch fermentation optimization using turbidity measurement system, *Escherichia coli, See Escherichia coli* fed-batch fermentation optimization using turbidity measurement system
Fed-batch mode of fermentation processes, advantages, 155,157
Fermentation
biosensor for on-line penicillin monitoring during production, 110–114
recombinant microorganisms, optimization of cultivation conditions, 155
Fermentation processes, modes of operation, 155
Ferrocene derivatives, use as sensor for subnanomolar concentrations, 70–80
Fiber-optic biosensor use in environmental monitoring
assay configuration, 35,37
biosensor probe, 34–36
evanescent wave, 34
fiber-optic fluorometer, 35,36*f*
groundwater contaminant detection, 40–41
total internal reflection, 34,36*f*
whole bacteria detection, 37–40
Fiber-optic fluorometer
function, 35
optical path, 35,36*f*
Flexible software system, use for flow injection immunoassay system automation, 165–174
Flow cytometry, applications, 107
Flow injection analysis, advances, 91–92
Flow injection analysis system(s)
penicillin monitoring, 110–114
programs for automation, 133,135
Flow injection analysis system supervision, multichannel, expert system, 133–142
Flow injection immunoassay(s)
application, 166
automation, 165–166
description, 165
Flow injection immunoassay system automation using flexible software system
computer system, 167,169,170*f*
experimental description, 166
heterogeneous system
control using CAFCA, 173,174*f*
procedure, 167,168*f*
homogeneous system
control using CAFCA, 171–172,174*f*
procedure, 166–167,168*f*
Fluorescence detection, alternative to electrochemical sensing, 100
Fluorescence lifetime measurement, description, 100–102*f*
Fluorescence resonance energy transfer, use in pH sensing, 105,106*f*
Fluorometry, lifetime-based phase, optical measurement of bioprocess and clinical analytes, 99–108
Fructose dehydrogenase, electroenzymatic sensing of fructose, 82–86
Fructose sensor, importance, 82
Fructose using fructose dehydrogenase immobilized in self-assembled monolayer on gold, electroenzymatic sensing, 82–86

G

Gluconobacter sp. fructose dehydrogenase, use as fructose sensor, 82
Glucose channel, expert system for multichannel flow injection analysis system supervision, 139–141
Glucose dehydrogenase, use as sensor for subnanomolar concentrations, 70–80
Glutamine and asparagine in aqueous media by near-IR spectroscopy, selective measurement, 116–130

Glutamine selective measurement in aqueous media by near-IR spectroscopy
absorption bands, 118–119
apparatus, 117
calibration models, 123,128–130
experimental description, 117,118
Fourier filtering effect on partial least-squares regression, 122–124*t*,127*f*
individual absorption spectra, 117,120*f*
partial least-squares regression, 119–122,124–126
Gold, electroenzymatic sensing of fructose using fructose dehydrogenase immobilized in self-assembled monolayer, 82–86
Groundwater contaminants, detection, using competitive binding assay, 40–41

H

Half-width, role in expert system for multichannel flow injection analysis system surpervision, 139,140*f*
High-performance LC, advances, 91–92
Human immunoglobulin G, use of acoustic wave device as LC detector, 9–17
Hybrid process modeling for advanced process state estimation, prediction, and control
application, 149,151–154
example of balanced equation system, 145,147
implementation, 149,150*f*
Monod-type approach, 147–148
need, 144
neural network approach, 147–148
structure, 144–149
Hybridoma technology, description, 29
Hydrogen peroxide measurement by reduction at low-potential, chemically modified electrode, 61–68

I

Image analysis, advances, 89–90
Imidazolinone herbicides, biosensor selectivity, 28*f*,29
Immunoassay(s)
concept, 19
design requirements, 21
Immunoassay system automation using flexible software system, flow injection, *See* Flow injection immunoassay system automation using flexible software system
Immunoglobulin G, sandwich immunoassay, 77,79*f*
Immunoreaction indication, enzyme sensors, 77
Immunosensor
antibody selectivity for the analyte, 25–29
cross-reactivity of constituents in chemical mixture, 25
detection limits, 22,25
measurement of antibody affinity for antigen, 20*f*,22
regeneration, 24–26*f*
sensitivity, 22
specificity, 22,23*f*
Intercalation
definition, 44
interaction with DNA, 44–45
Intercalative interaction of dyes with DNA, characterization, 45

L

Laccase, use as sensor for subnanomolar concentrations, 70–80
Lifetime-based phase fluorometry, optical measurement of bioprocess and clinical analytes, 99–108
Liquid chromatographic detector, acoustic wave device, 9–17
Low-potential, chemically modified electrode for hydrogen peroxide measurement by reduction, 61–68

M

Mammalian cell culture, production-scale, for hybrid process modeling of advanced process state estimation, prediction, and control, 144–154

Microbial receptors, examples, 3
Monitoring, bioprocesses, advances, 88–94
Monoclonal antibodies, production, 29–31
Multichannel flow injection analysis system supervision, expert system, 133–142

N

Naphthalene, detection and measurement by DNA intercalation, 44–58
Near-IR spectroscopy
 development as continuous monitors for bioreactors, 116–117
 glutamine and asparagine selective measurement in aqueous media, 116–130
Nitrobenzene, detection and measurement by DNA intercalation, 44–58
Norepinephrine detection, enzyme sensors, 75,76*f*

O

1,2,3,4,5,6,7,8-Octahydronaphthalene, detection and measurement by DNA intercalation, 44–58
Off-gas analysis, advances, 92–93
On-line measurements
 flow injection analysis, 91–92
 high-performance LC, 91–92
 image analysis, 90–91
 off-gas analysis, 92–93
 optical density probes, 89–90
On-line penicillin monitoring during production by fermentation, biosensor, 110–114
Optical measurement of bioprocess and clinical analytes using lifetime-based phase fluorometry
 fluorescence lifetime measurement, 100*f*–102
 future, 107–108
 oxygen sensor, 101–104*f*
 pH–pCO_2 sensor, 103,105–107
Optimization, *Escherichia coli* fed-batch fermentation using turbidity measurement system, 155–164
Oxygen sensor, use of lifetime-based phase fluorometry, 101–104*f*
Oxygen uptake rate
 on-line off-gas analysis for determination, 92–93
 recirculation method, 93
 stationary method, 93

P

Parathion, detection and measurement by DNA intercalation, 44–58
Penicillin biosensor monitoring during production by fermentation
 automation procedure, 111–112
 calibration curve, 113
 comparison to high-performance LC method, 113,114*f*
 detection cell, 111,112*f*
 flow injection system, 112
 importance of measurement, 113
 penicillin concentration vs. fermentation time, 113,114*f*
 response time, 114
 stability, 114
Pentachlorophenol, detection and measurement by DNA intercalation, 44–58
Pesticides, detection, 40–41
pH sensor, use of lifetime-based fluorometry, 103,105–107
Phagemids, description, 30
Phase fluorometry, lifetime-based, optical measurement of bioprocess and clinical analytes, 99–108
Phase-modulation fluorometry, description, 100–102*f*
Phenanthrene, detection and measurement by DNA intercalation, 44–58
Polarization, calculation, 48

Poly(anilinomethylferrocene)-modified glassy carbon electrode for hydrogen peroxide measurement by reduction at low potential
analytical characterization, 66,67*f*
application, 67–68
construction, 62–63
effect of number of electrochemical polymerization scans, 63,64*f*
effect of pH, 63,65–66
experimental description, 62–63
hydrodynamic voltammetry, 63,64*f*
interference, 66–67
Polyaromatic carcinogen detection and measurement by DNA intercalation
carcinogen measurement by acridine orange displacement, 49,51–53
carcinogenicity vs. detection limit, 57–58
criterion for competition against acridine orange, 52–53,55*t*
effect of molecular weight, 54–57
effect of size, 53–54,55*t*
experimental description, 45–48
observed polarization change mechanism, 48–51*f*
previous studies, 45
system polarization response determination procedure, 47–48
Polychlorobiphenols, detection, 40–41
Postcolumn on-line detectors
surface transverse wave device, 11
two-cell reactors, 11
Process(es), importance of monitoring and control, 116
Process control, monitoring using biomolecular recognition, 4–5
Process state estimation, prediction, and control exemplified at production-scale mammalian cell culture, hybrid process modeling, 144–154
Production process, control prerequisites, 144
Production-scale mammalian cell culture, hybrid process modeling for advanced process state estimation, prediction, and control, 144–154
Proflavin, use in DNA intercalation, 44–58
Protease stopped-flow channel, expert system for multichannel flow injection analysis system supervision, 141,142*f*
Protein folding properties, dependence on cultivation conditions of host organism, 144

R

Radioimmunoassay, applications, 19
Recombinant microbial and cell culture technology, development, 88
Redox label immunoassay, enzyme sensors, 80
Reduction at low-potential, chemically modified electrode for hydrogen peroxide measurement, 61–68
Residence time, role in expert system for multichannel flow injection analysis system supervision, 139,140*f*
Respiration quotient, on-line off-gas analysis for determination, 92–93

S

Sandwich assay
procedure, 35
whole bacteria detection, 37–40
Sandwich immunoassay for immunoglobulin G, enzyme sensors, 77,79*f*
Selective detector, examples, 9
Selective measurement, glutamine and asparagine in aqueous media by near-IR spectroscopy, 116–130
Selectivity
antibody for analyte, 25–29
detection using immunosensors, 19–31
Self-assembled monolayer on gold, electroenzymatic sensing of fructose using fructose dehydrogenase, 82–86
Sensitivity, immunosensor, 22
Sensor system, ideality, 33
Sensorgram, description, 13,15
Sol-gel, use in pH sensor, 105,107
Specificity, immunosensor, 22,23*f*
Subnanomolar concentrations, enzyme sensors, 70–80

Supervision, expert system for multichannel flow injection analysis system, 133–142
Surface acoustic wave devices, *See* Acoustic wave device as LC detector
System polarization response, calculation, 47

T

Time, role in electroenzymatic sensing of fructose using fructose dehydrogenase immobilized in self-assembled monolayer on gold, 85
Time-resolved fluorescence, 100–102*f*
Total internal reflection, description, 34,36*f*
Trichloroethylene, degradation by *Burkholderia cepacia* G4 5223–PR1, 37–40
Trinitrotoluene, detection, 40–41
Turbidity measurement system, *Escherichia coli* fed-batch fermentation optimization, 155–164

U

Universal detector, example, 9

W

Whole bacteria, detection using sandwich assay, 37–40

Production: Charlotte McNaughton & Amie Jackowski
Indexing: Deborah H. Steiner
Acquisition: Michelle D. Althuis
Cover design: Amy Hayes

Printed and bound by Maple Press, York, PA